Primary and Secondary Brain Stem Lesions

By

G. Csécsei, O. Hoffmann, N. Klug, A. Laun,
R. Schönmayr, J. Zierski

Acta Neurochirurgica
Supplementum 40

Springer-Verlag Wien New York

Dr. György Csécsei
Dr. Oskar Hoffmann
Professor Dr. Norfrid Klug
Dr. Albrecht Laun
Dr. Robert Schönmayr
Professor Dr. Jan Zierski

Department of Neurosurgery, University of Giessen, Federal Republic of Germany

With 80 Figures

Library of Congress Cataloging-in-Publication Data. Primary and secondary brain stem lesions. (Acta neurochirurgica. Supplementum, ISSN 0065-1419; 40). Includes bibliographies. 1. Brain stem—Diseases. 2. Brain stem—Hemorrhage. 3. Brain—Tumors—Complications and sequelae. 4. Cerebrospinal fluid pressure. I. Csécsei, G. (György), 1948– . II. Series. [DNLM: 1. Brain Diseases—physiopathology. 2. Brain Stem—physiopathology. W1 AC8661 v. 40/WL 310 P952] RC394.B7P74 1987. 616.8. 87-31089

ISSN 0065-1419
ISBN-13:978-3-7091-8943-6   e-ISBN-13:978-3-7091-8941-2
DOI: 10.1007/978-3-7091-8941-2

Foreword

Lesions of the brain stem and related disorders of autonomous regulation systems have been the "Leitmotiv" of the scientific work in the Neurosurgical Department of the Giessen University under the leadership of the late Professor Hans Werner Pia. Some of the results have already been published in supplement volumes of Acta Neurochirurgica. The first one of these was Vol. 4 with the monograph written by H. W. Pia on "Die Schädigungen des Hirnstammes bei den raumfordernden Prozessen des Gehirns". Vol. 19 deals with central disorders of temperature regulation, written by G. Lausberg, followed by Vol. 20 with R. Lorenz's monograph on the effects of intracranial space-occupying lesions on blood pressure and heart rate.

Shortly before his death Hans Werner Pia had asked me to combine and publish in this series of supplement volumes of Acta Neurochirurgica another five papers by his co-workers, also related to brain stem lesions and also to cerebral blood flow and CSF dynamics. The result is this volume which contains work dealing with the CT-findings of cerebral mass displacements and their clinical correlations (Schönmayr), with other lesions of the brain stem (Laun), with the blink reflex and acoustic evoked potentials in brain stem lesions (Klug and Csécsei), with blood flow in brain structures during increased ICP (Zierski) and with the description of a mathematical model for analysis and simulation of the haemodynamics of intracranial CSF (Hoffmann).

The papers not only present valuable scientific information but at the same time reflect the scientific activity of Hans Werner Pia and his team. As a fitting memorial they are dedicated to him.

F. Loew, Homburg/Saar, Federal Republic of Germany

Contents

Electrically Elicited Blink Reflex and Early Acoustic Evoked Potentials in Circumscribed and Diffuse Brain Stem Lesions. By N. Klug and G. Csécsei   57

Acta Neurochirurgica, Suppl. 40, 1–27 (1987)

Cerebral Mass Displacements
Part I: Cisternal Hernia in Intracranial Tumours in the Computer Tomogram
Part II: Clinical Findings in Primary and Secondary Brain Stem Lesions

Robert Schönmayr

Department of Neurosurgery, University of Giessen, Federal Republic of Germany

Contents

Part I

Cisternal Hernia in Intracranial Tumours in the Computer Tomogram

Introduction

Arnold (1894) and Chiari (1896) described the displacement of the cerebellar tonsils through the foramen magnum into the initial part of the spinal canal. However, they related this observation exclusively to the simultaneous occurrence of malformations. Alquier (1905) and Henschen (1910) reported such tonsillar alterations in intracranial tumours. Cushing (1917 and 1929) recognized the connection with the increase of intracranial pressure and coined the term "cerebellar pressure cone".

A first more extensive description of alterations in the form of the brain at bone and dura projections in raised intracranial pressure was made by Meyer (1920). Besides the mass displacement taking place under the falx, he described above all the protrusion of the medial parts of the temporal lobe over the tentorial margin, and chose the expression "hernia" for these processes.

In his studies with Stroescu (1934) and Hasenjäger (1937), Spatz presented his comprehensive concept of mass displacements of the brain in space-occupying processes. The significance of special anatomical features of the cisterns and the alterations of the parts of the brain pressed into the cisternal closure due to swelling was emphasized.

Vincent, David, Thiébaut and Rappoport (1930, 1936) had investigated connections between what they termed "temporal pressure cone" and specific clinical states. Temporal herniation and the clinical symptoms caused by it were also studied by Bailey (1933), van Gehuchten (1933), Bannwarth (1935), Olivecrona (1936), Jefferson (1937) Cairns (1937), Mansuy (1937), and Le Beau (1938).

The essential foundations of the view of "alterations in the form of the brain in space-occupying processes" which is still valid today was laid by Riessner and Zülch (1939) and Tönnis, Riessner and Zülch (1940) with their studies on this topic. They described the passive displacements of the brain mass owing to increase of volume and pressure and its principles in relation to the nature, localization and concomitant circumstances of the space-occupying process. Pia (1956) gave a synopsis of the mass displacements in raised brain pressure and in particular in cisternal hernia together with a comprehensive description of their clinical effects, above all on the brain stem.

All the investigations mentioned are based on an extensive autopsy material and were also supported by intraoperative biopsy observations. However, a systematic investigation of cisternal behavior in space-occupying processes intra vitam was only made possible by computer tomography. The majority of the cisterns described in pathological anatomical investigations and recognized as significant are also to be regularly identified on computer tomographic recordings. Difficulties result only in the cisterns of the posterior cranial fossa so long as they were still imaged with instruments of the first and second generation. Because of frequent artefacts based on the technique, above all the cerebellomedullary cistern, the median and lateral parts of the pontocerebellar and pontomedullary cistern can only be appraised in rare cases. Only the images obtained with instruments of the most recent generation permit unequivocal appraisal of these cisterns.

The objective of my investigation was to observe the behavior of the cisterns in intracranial space-occupying processes in intracranial tumours as an example and to establish how individual cisterns react to tumours of a specific size and localization, and to what extent and in what form they are affected by alterations.

Conversely, the question was also to be clarified as to whether depending on their location, specific sequence or characteristic combinations tumours lead to an impairment of the cisterns.

Materials and Methods

A total of 800 CT recordings were used for evaluation. These were computer tomograms of 289 patients with intracranial tumours, of these 59 patients with tumours of the posterior cranial fossa. Up to ten CT were evaluated per patient. The person-related data of the patient, data on the diagnosis, time course and outcome of the disease were recorded on a survey form. The documentation of the computer tomographic findings was made in the form of a description of the alterations which can be discerned in the CT. Much space is taken up by the documentation of possible signs of mass displacement. Each individual region of the internal and external CSF spaces was appraised with regard to compression or enlargement and displacement. After exclusion of inadequately imaged CT or CT which could not be evaluated owing to artefacts, each individual cistern was appraised with regard to its form, width and the nature of the impairment. Since the CT recordings were done by a machine of the second generation, the cisterns of the cerebellopontine angle and the cerebellomedullar cistern could not be visualized in a sufficient number for evaluation owing to artefacts.

The data obtained in this way were transferred onto a marking sheet which could be read by machine for further processing, transferred to punch cards and these were read onto magnetic tape. For evaluation, the behavior of each individual cistern was now considered taking into account the location of the tumour. Besides the cistern width, in the case of the paired cisterns, their symmetry was also evaluated.

CT recordings in which there was a dilatation of the inner CSF spaces with simultaneous narrowing of the external CSF spaces were considered separately. These recordings were listed separately from the remaining localizations in the tables under the term "hydrocephalus".

In order to be able to evaluate the behavior of the cisternal spaces in space-occupying processes of different sizes, a distinction was made between "large" and "small" processes on the basis of the degree of displacement of the midline structures. The midline displacement is the result of all space-occupying components taken together, i.e. tumour + perifocal oedema. A small tumour with large oedema may bring about a more pronounced mass displacement than a major tumour with slight perifocal oedema. Since the sum of all intracranial space-occupying components is always decisive for the behaviour of cisterns, the degree of midline displacement was taken as a yardstick. The term "tumour" is not used in its histopathological sense, but with its original Latin meaning as "swelling, accumulation".

The problem of preoperative and postoperative CT recordings is also solved by the definition of "tumour size" as the extent of the midline displacement. In consequence of the postoperative decline of mass displacement with accompanying stepwise normalization of cisternal findings, the CT in which distinct mass effects are still to be seen, are to be found amongst the "large tumours", whereas with increasing regression of the displacements the corresponding CT are classified under the "small tumours". Complete normal computer tomograms in which no space-occupying process was to be detected were not considered.

Results

Influence of Tumour Location on Alterations of the Cisterns

1. Tumours of the Frontal Lobe (Figs. 1, 2, 3)

Among the tumours of the frontal lobes, three groups can be distinguished with regard to cisternal alterations they induce:

1. frontal and frontomedial
2. frontobasal
3. frontolateral.

The proportion of CT with major mass displacement to the opposite side is of the same order of magnitude in the frontal tumours (38%) as in the occipital tumours (37%). Since in all other localizations with the exception of frontobasal the proportion of major mass displacements is about 50%, special anatomical features become evident from this.

Thus tumours near to the frontal and occipital pole develop their thrust effect initially in the sagittal direction supported by the resistance of the anterior or posterior falx and lead to mass displacement only after using up the reserve spaces and general rise of pressure on the side of the affected hemisphere.

On the other hand, the further away a space-occupying process is from the cerebral poles the greater is its thrust effect in the coronary plane. This can be directly converted into mass displacement.

Finally, for the cisternal behaviour as a whole the general mass displacement is a more important factor than the size of the space-occupying process, which in turn results from several factors (tumour, perifocal oedema, CSF congestion, possibly haemorrhage). However, with the exception of midline tumours, the size of which is not adequately manifested in the lateral displacement of the ventricles, there is a correlation between the midline displacement and the size of the space-occupying process under the restrictions mentioned (here: tumour and reaction of the surroundings). To simplify, tumours which are accompanied by an unequivocal midline displacement are designated as "large" and those without such a displacement are designated as "small".

1.1. Frontal Tumours (Fig. 1)

In their effect on the cisterns detectable in the CT, owing to their close vicinity, the frontal tumours most frequently show a narrowing or occlusion of the anterior interhemispheric spaces. The high involvement of the Sylvian cisterns and the posterior interhemispheric fissure characterizes these as "reserve spaces" from which the CSF is pressed out as from the entire subarachnoid space. The high rate of asymmetry of these alterations between 71% and 77%, which only decreases even in the large tumours with a greater mass displacement (64% to 69%) shows that this mainly takes place unilaterally or at least predominantly on one side.

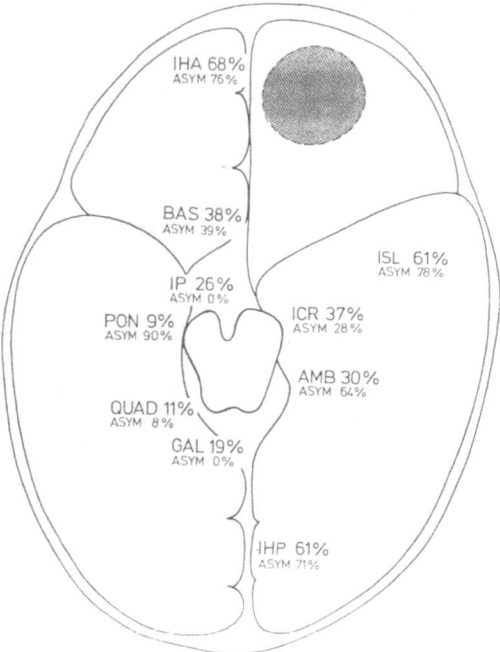

Fig. 1. Frontal tumours

The basal cistern and the intercrural cistern are affected with the next highest frequency (38% and 37%). Their compression shows a very much lower side difference (39% and 28% respectively) and thus reveals the pushing force from the front, which is also indicated by the more frequent narrowing of the interpeduncular cisterns.

The ambient cistern behaves highly asymmetrically. The frequent homolateral dilatation shows the displacement of the brain stem towards the opposite side, whereas the high proportion of symmetrical bilateral narrowing in simultaneous minimal asymmetry of the quadrigeminal cistern indicate a simultaneously occurring dorsal displacement. The involvement of the cistern of Galen at a level of 19% also points in the same direction.

The pontine cistern, hardly impaired by the frontal tumours, is mainly dilated, also indicating a displacement of the brain stem away from the clivus. The high asymmetry which is to be observed here arises from the simultaneous lateral displacement.

To summarize, frontal tumours initially compress the anterior interhemispheric and Sylvian cisterns in the immediate spatial vicinity. Next, the basal cistern (in large tumours, also the interpeduncular cistern) are occluded, but at the same time, the intercrural cisterns are also compressed from the front.

The remaining cisterns reflect the displacement of the brain stem to dorsal and to the opposite side.

1.2. Frontobasal Tumours (Fig. 2)

As typcial basal tumours mostly situated near or in the midline, they form a group of their own with quite characteristic effects on the cisterns. They cause only slight side differences. The highest rate of asymmetry (33%) reaches the anterior interhemispheric fissure, since here the falx mostly prevents or alleviates a direct compression of the contralateral cisterns even in tumours situated quite close to the midline. Otherwise, the symmetry is between 0% and 25%.

By far the most frequent narrowing, obstruction or (to put it better) in many cases tamponade occur in the basal and interpenduncular cisterns. The latter are only effective when the basal cistern is already filled by the tumour. This growth of tumours into the basal cisternal cavities also explains the symmetrical compression of the intercrural cisterns.

The planation of the CSF spaces over the convexity (i.e. also of the Sylvian cistern and the interhemispheric cistern) only takes place afterwards.

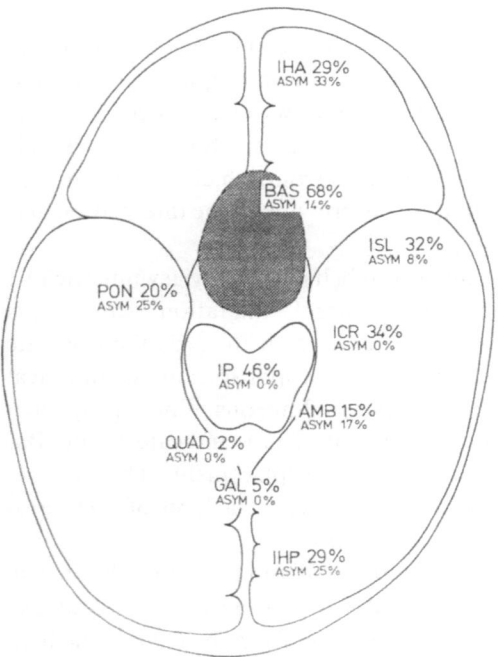

Fig. 2. Frontobasal tumours

The ring of ambient, quadrigeminal and Galen cisterns is also symmetrically narrowed by the dorsal displacement of the brain stem. The rate of involvement of these cisterns is less than in the frontal tumours, however, possibly the basal cisterns afford more abundant space for expansion of the tumours.

The pontine cistern reveals two mechanisms: either it is also reached and filled by the tumour, or it is enlarged by the displacement of the brain stem away from the clivus.

1.3. Frontolateral Tumours (Fig. 3)

These take up an intermediate position between the frontal and the temporal tumours. They also primarily lead to compression of the CSF spaces over the convexity, the Sylvian cistern (76%) and the anterior and posterior interhemispheric spaces (75% and 71%).

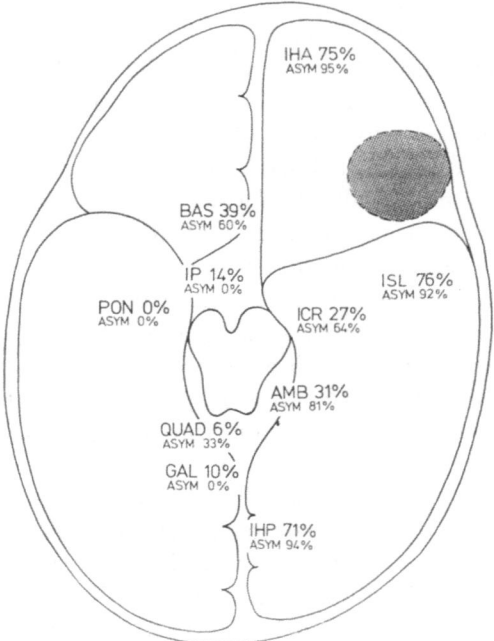

Fig. 3. Frontolateral tumours

In all the cisterns mentioned, the alterations are asymmetrical in more than 90%. In the small tumours which do not elicit any appreciable displacement of the midline structures, the asymmetry is indeed 100% in all these cisterns. The major involvement of the Sylvian cistern appears to be a manifestation of the herniation of the frontal lobe over the wing of the sphenoid, although this process is hardly visualized once in CT, probably because of the axial orientation of the tomograms.

The basal cistern which is affected in the same order of magnitude (39%) as in the frontal tumours shows a greater side difference (60%).

The intercrural cistern with 64% asymmetry is preferentially compressed on the tumour side.

The ambient cistern with a high proportion of homolateral dilatation and 81% asymmetry reveals the lateral displacement of the brain stem more clearly than in the frontal tumours. The closely situated cisterns of the quadrigeminal plate and Galen are more rarely narrowed than in frontal tumours and thus indicate the lower dorsal displacement of the brain stem.

1.4. Summary

If the alterations are summarized, it is shown that the cisterns situated closer are once more preferentially compressed, the Sylvian cistern being involved to a greater extent with the more lateral position of the tumours. If the compression of the basal and intercrural cistern was mainly symmetrical in the frontal tumours, it is very much more pronounced on the tumour side in the frontolateral tumours. With the greater compression on one side, above all of the intercrural cistern, the brain stem is displaced almost only to lateral, but hardly at all to dorsal. This can be seen from the high asymmetry of the lateral ambient regions and the slight narrowing of the mesencephalic tectum cistern and the cistern of Galen. No assessment of the cerebellomedullary cistern had been possible due to the already mentioned technical reasons.

2. Tumours of the Temporal Lobe (Figs. 4 and 5)

The basal tumours are to be separated from the temporally situated tumours, since they have other effects on the cistern system. In their behaviour towards the cisterns, they display similarities with the temporal tumours, but also with frontobasal tumours.

2.1. Temporal Tumours (Fig. 4)

Compression of the Sylvian cistern (70%) takes first place, followed by that of the interhemispheric fissure, which is equally affected to the extent of 67% at the front and at the back.

The side differences are pronounced (75%–82%), even if not as high as in frontolateral tumours owing to the higher proportion of cisterns completely obstructed on both sides.

More frequent than the narrowing of the ambient cistern which is very pronounced on one side, is its one-sided dilatation, a manifestation of the lateral displacement of the brain stem in the tentorial aperture.

In contrast to this, unilateral cisternal dilatation hardly occurs in the intercrural cistern, in which compression predominates. This mostly affects both sides and is more pronounced on one side in 58%. It appears that its compression causes the displacement of the brain stem to the opposite side and the dilatation of the ambient cistern of the same side associated with this.

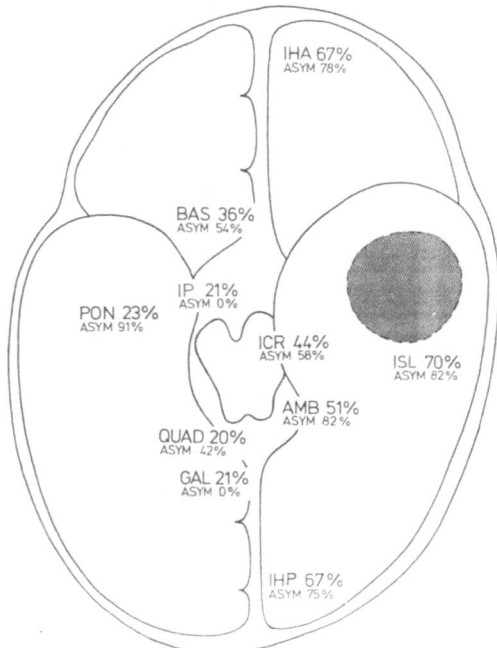

Fig. 4. Temporal tumours

The narrowing of the basal cistern (36%) shows a side difference in 54%.

The quadrigeminal cistern and the cistern of Galen are more frequently narrowed or obstructed in temporal tumours than in frontolateral tumours. This is due to the fact that it is now no longer the dorsal displacement of the brain stem which brings about the narrowing as in the frontal tumours, but the herniation of the hippocampal gyrus increasingly also involves the posterior parts of the tentorial notch and thus the spaces of these two cisterns. This results from the behaviour of the ambient cistern, which simultaneously reveals a homolateral narrowing. Only a small part of the side differences of the mesencephalic tectum derive from homolateral dilatation.

With its pronounced asymmetry and increased unilateral dilatation, the pontine cistern provides above all a picture of the lateral displacement of the brain stem. The bilateral dilatation which can be detected in a few cases points to a sagittal component with displacement of the brain stem from the clivus.

To summarize, two, partly competing mechanisms become evident in temporal tumours: on the one hand, herniation via the tentorial margin with narrowing or obstruction of the homolateral perimesencephalic cisterns. On the other hand, the lateral displacement of the brain stem with dilatation of the cisternal ring on the side of the tumour. Which process is manifested doubtlessly depends on the location of the tumour in the temporal lobe. However, for a further breakdown into tumours of the temporal pole, of the middle and posterior temporal lobe, our numbers were too small in order to produce results which could be utilized.

2.2. Temporobasal Tumours (Fig. 5)

Sylvian and interhemispheric cisterns are also the first to be affected in these tumours, even if in a reduced incidence (54%–59%). The intercrural cistern comes to the fore (56%). It is often directly compressed by the tumour (tamponade). The contralateral cistern is frequently also narrowed, which is why the asymmetry keeps within bounds (48%).

Basal and interpeduncular cisterns are involved to an especially great extent, reflecting the basal position of the tumours.

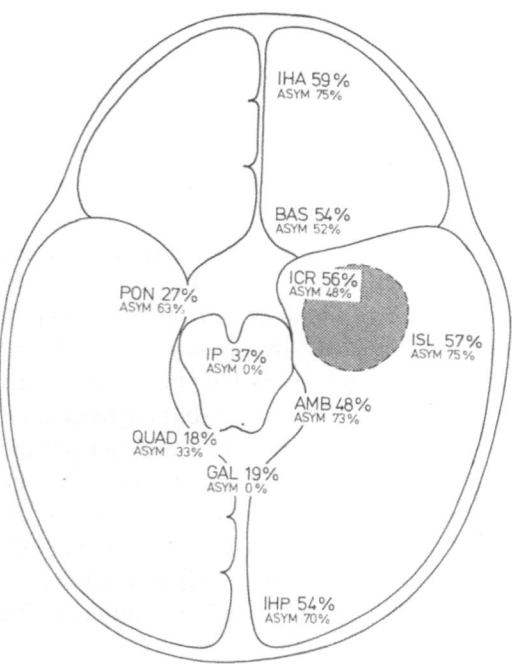

Fig. 5. Temporobasal tumours

The perimesencephalic cistern ring shows highly asymmetric behaviour with slight predominance of the alterations based on the brain stem displacement compared to the alteration caused by temporal herniation.

The pontine cistern is either tamponaded by the tumour or unilaterally dilated by lateral displacement of the brain stem, especially in a tumour situated farther cranially.

If the cistern alterations in temporal basal tumours are summarized, there is a decline of the involvement

and the rate of asymmetry of the interhemispheric spaces and the Sylvian cistern compared to the temporal and frontolateral cisterns. On the other hand, they show pronounced compression of the intercrural cistern.

3. Tumours of the Parietal Lobe (Fig. 6)

The cisterns of the Sylvian and the interhemispheric fissure are preferentially affected. However, a more occipital distribution of the compression is now already shown. Thus the occipital segment of the interhemispheric subarachnoid spaces is by far the most frequently involved (92%).

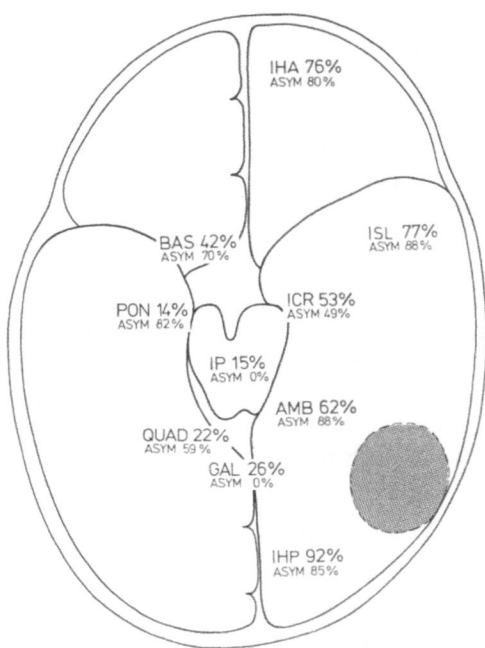

Fig. 6. Parietal tumours

The major effect on the ambient cisterns (62%) is explained from the immediate vicinity. There the mainly unilateral, homolateral compression of the cistern predominates, an unmistakable sign of herniation of the hippocampal gyrus. However, the proportion of homolateral dilatation with and without compression of the contralateral cistern is also high. Owing to the high rate of asymmetry of 88%, the mass displacement also proves to be a major factor for cistern narrowing in parietal tumours.

The pronounced side differences also affect the quadrigeminal cistern, the alterations of which reach a peak in parietal tumours (22%). Whether small tumours lead to dilatation of the cistern via the displacement of the brain stem, large tumours compress the cistern by herniation.

The cistern of the vein of Galen is also most frequently compressed by parietal tumours. In contrast to the quadrigemina, it is already compressed by smaller tumours which first of all touch it owing to the spatial relationship.

Next to the basal tumours, the parietal tumours exceed all other lobar tumours in their effect on the basal cisterns. At frist glance, the high proportion of compression of the intercrural cistern, which only presents side differences in about half of the cases, might indicate a ventral displacement of the brain stem in a thrust effect emanating from dorsal and lateral. Since that at the same time no narrowing is to be detected at the pontine cistern, this mechanism becomes implausible.

This is explained by the craniocaudal displacement of the brain stem, as indicated by the high rate of narrowing of the basal cistern. By entry of the crura cerebri into the tentorial notch, a bilateral mainly symmetrical narrowing of the intercrural cisterns arises. A compression of the pontine cistern is not discerned in this process. On the contrary, an increased distance from the clivus is simulated by the descent of the interpeduncular cistern (14%).

The asymmetry of the alterations affecting the basal cistern (70%) and pontine cistern (82%) shows that besides the axial displacement, a lateral component also acts on the brain stem.

To summarize: as a mirror image of the frontal tumours, the narrowing initially takes place in the posterior interhemispheric spaces of the vicinity and then in the external CSF spaces via the overall convexity.

The herniation of brain tissue beyond the tentorium margin concerns above all the dorsal side of the perimesencephalic cistern ring (quadrigeminae, Galeni). On the other hand, its ventral parts, the basal cisterns and the pontine cistern show the characteristics of axial or craniocaudal displacement of the brain stem (champagne cork phenomenon according to Spatz). It is superimposed by a second, lateral component, as a manifestation of the mass displacement.

4. Tumours of the Occipital Lobe (Fig. 7)

Like the parietal tumours, they frist of all compress the adjacent posterior interhemispheric spaces (82%), then the Sylvian and the anterior parts of the interhemispheric fissure (61% each).

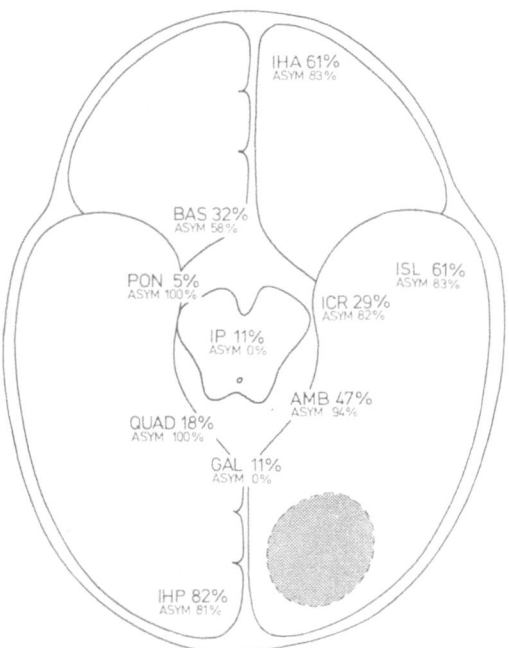

Fig. 7. Occipital tumours

In the ambient cistern and the quadrigeminal cistern, the alterations are characteristic for a displacement of the brain stem. The pronounced asymmetry in both derives from the large number of homolateral dilatations and contralateral compressions. Direct narrowing by herniation does not occur in the ambient cistern and is manifested in the quadrigemina in only 5% of the CT. The cistern of Galen is very much more rarely compressed (11%) than in parietal or temporal tumours. The appreciable side differences in the intercrural cistern (82%), characterized by homolateral dilatation and contralateral narrowing in one third, indicate that the midline displacement is the crucial factor. The basal and interpeduncular cistern are only affected to a slight extent by small tumours (13% and 4% respectively). Only tumours with a distinct mass displacement to the opposite side impair the two systems more (64% and 21% respectively). With its minimal involvement (5%), the pontine cistern is the only one to show signs of lateral brain stem displacement.

To summarize, occipital tumours exert direct compression only on the posterior interhemispheric fissure, and on the other cisterns they act mainly by the mass displacement resulting from the general unilateral increase in volume. The herniation at the tentorial notch plays only a minor role here. There are no signs of ventral displacement.

5. Tumours of the Basal Ganglia (Fig. 8)

These tumours, among which thalamic tumours are also to be found, cause alterations which markedly deviate from those of tumours of other locations. However, their behaviour with regard to the cisterns can only be interpreted with reservations because of their small number (n = 19).

The Sylvian cistern is affected in all investigations. It is directly compressed by the tumours, only unilaterally or predominantly on one side. The proportion of homolateral dilatation is a manifestation of the post-operative defect. The anterior and posterior region of the interhemispheric fissure follow with 79% and 74% respectively. They thus reveal a somewhat greater frontal narrowing. A bilateral narrowing of the cisterns only occurs with appreciable mass displacement and is always more pronounced on the one side.

The next most frequent involvement (37%), exclusively in the form of compression, is displayed by the intercrural cistern. The marked asymmetry (71%) with more intense narrowing on the tumour side corresponds to uncus hernia.

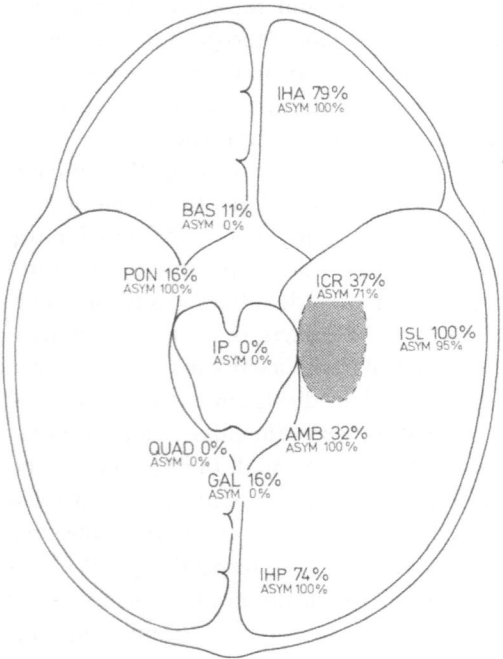

Fig. 8. Basal ganglia tumours

On the other hand, the ambient cistern presents signs of lateral brain stem displacement in 100% asymmetry: homolateral dilatation and contralateral narrowing. Whereas the quadrigeminal cistern does not show any involvement, the cistern of Galen is com-

pressed in 16% of the CT. In connection with the exclusively contralateral narrowing of the ambient cistern, which indicates an appreciable hernia of the posterior regions of the hippocampus, a sagittal component of the mass displacement is to be assumed as the cause of the narrowing of the cistern of Galen.

The involvement of the basal cistern (11%) is slight, and the interpeduncular cistern is not affected at all. This behaviour is difficult to interpret since a greater narrowing also of the basal cisterns would be expected in view of the position of the tumors near to the midline. The pontine cistern is not compressed in any case. Its alterations consist exclusively of unilateral dilatation (16%). This is a sign of displacement of the brain stem.

To summarize, the basal ganglia tumours cause cistern alterations mainly via the general increase in volume of the affected hemisphere with abolition of the external CSF cavities, uncus hernia and lateral displacement of the brain stem. On the other hand, a direct compression, e.g. of the basal cisterns located near to them does not occur.

6. Tumours of the Posterior Cranial Fossa

6.1. Midline Tumours (Fig. 9)

They comprise the tumours of the median cerebellar structures, the fourth ventricle and the brain stem. The number of investigations 24 is small, above all because investigations in which an occlusion hydrocephalus could be detected were considered separately.

The most frequent alterations are found in the ambient cistern (71%) and intercrural cistern (58%). In a considerable proportion, they are asymmetric (59% and 79% respectively). The processes situated near to the midline are evidently also due to the side differences which can be discerned in the paired cisterns.

The median, dorsal cisterns of the mesencephalic ring are frequently compressed (quadrigemina in 42%, Galeni in 38%), but show only slight side differences (10%).

This apparently contradictory involvement of the perimesencephalic cisterns can be explained by two mechanisms. On the one hand, the tumours situated in the brainstem lead to an increase in its volume, resulting in a narrowing of the cistern. This naturally occurs to a particular extent when the tumour is located in the midbrain itself. Even a slightly paramedian location of the tumour results in side differences in compression. A distension of the midbrain also leads to narrowing of the interpeduncular cistern (21%) and basal cistern

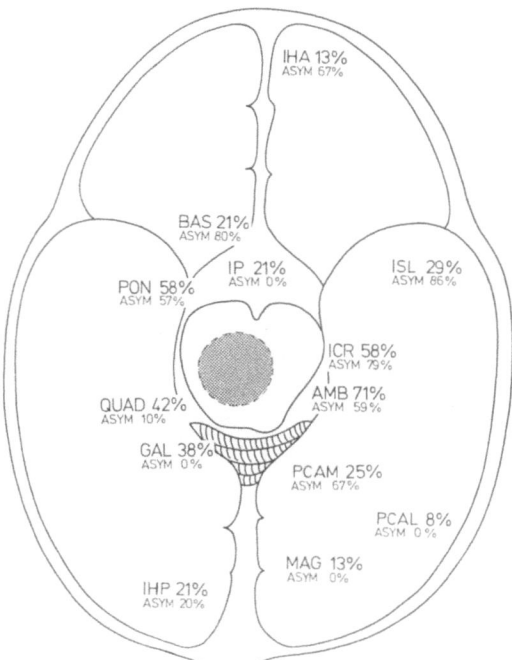

Fig. 9. Fossa posterior: midline tumours

(21%) and can also result in asymmetry of these cisterns (80%).

On the other hand, the tumours situated dorsal to the brain stem bring about its displacement to ventral, and thereby reduce the distance between the pons and the clivus, but also between the crura cerebri and the unci gyri hippocampalis. By the increase in volume in the cerebellum, there is frequently a simultaneous herniation of the apical vermicular process to cranial in the direction of the quadrigemina and Galen cisterns. Whereas the ventral displacement of the brain stem already results in the occurrence of a slight lateral component in a asymmetrical configuration of the perimesencephalic cisterns, the ascending cerebellar hernia brings about a symmetrical narrowing of the dorsal cisterns. In accordance with this mechanism, the pontine cistern shows narrowing in all midline tumours situated outside the brain stem. On the other hand, a small proportion of tumours localized within the brain stem is also able to bring about a unilateral dilatation of the cistern as sign of a lateral displacement component. The involvement of the remaining supratentorial cisterns (Sylvian, interhemispheric fissure) is slight, and is more pronounced occipitally.

The rate of compression of the medial and lateral cerebellopontine angle cisterns and the cerebellomedullary cistern cannot be assessed with sufficient certainty, since they could not be evaluated regularly because of frequent artefacts.

Summary: tumours of the midline of the posterior cranial fossa elicit above all cistern alterations in the sense of an "ascending" herniation: due to the infratentorial increase in volume, the apical cerebellar regions are displaced upward into the eaves of the cerebellar tent. The brain stem is displaced to ventral and confines the pontine cistern. As long as the perimesencephalic cisterns can still be identified, they very clearly indicate even tiny lateral displacements of the brain stem.

6.2. Tumours of the Cerebellar Hemispheres (Fig. 10)

In these tumours, in principle the same mechanism is manifested as in midline tumours of the posterior fossa. Firstly, the ventral displacement of the brain stem, secondly the cerebellar herniation through the tentorial notch to cranial. Once more, the brain stem displacement shows lateralization effects owing to the paramedian position of the tumours and the lateral component of the direction of thrust due to this, whereas the dorsal and mostly median hernia of the cerebellum preferentially takes place symmetrically.

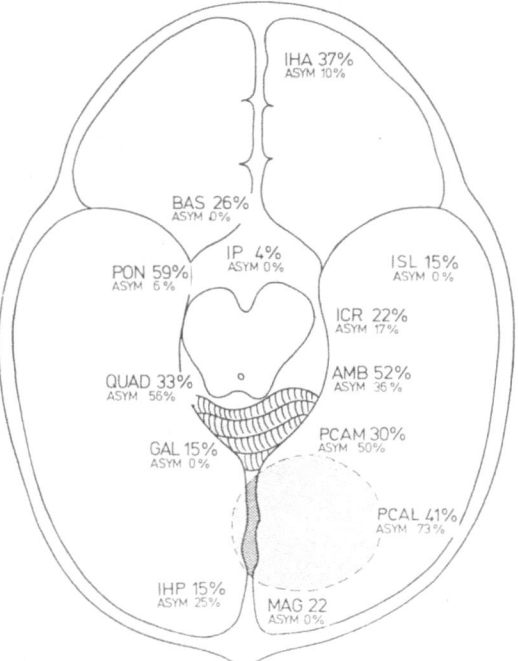

Fig. 10. Tumours of the cerebellar hemispheres

The side-different behavior of the perimesencephalic cisternal ring loses significance with increasing tumour size, since tumours with pronounced space-occupying effect finally lead to a diffuse blockade of the cisterns which no longer allows a side differentiation. Nevertheless, 36% side differences can be discerned in the ambient cistern, and even 56% in the quadrigeminal cistern. The intercrural cistern with mainly bilateral compression shows only 17% asymmetry, a consequence of the ventral displacement of the brain stem. There is otherwise slight asymmetry also in the other supratentorial cisterns. The same applies to the basal cistern. Compression of the latter and of the interpeduncular cistern is minimal. The pontine cistern is involved most frequently (59%), almost only symmetrically in these tumours.

In the evaluation of the other infratentorial cisterns, once more substantial reservations are to be made. Nevertheless, it is apparent that the lateral cerebellopontine angle cistern is affected to a greater extent than the medial cistern, and that side differences are manifested more distinctly here. In reality, the cerebellomedullary cistern is likely to be very much more frequently involved than shown here, since it was almost always narrowed or occluded in a few cases in which it could be evaluated.

To summarize, cerebellar hemispheric tumours largely resemble the midline tumours of the posterior fossa in their effect on the cisterns. In the CT, the "ascending" herniation is decisive.

Herniation of the cerebellar tonsils into the greater foramen, which is a mechanism more familiar from clinical, biopsy and autopsy findings, and which according to operative findings is doubtlessly very important, could not be detected in an adequate number in the available registrations owing to interference by artefacts. Information on this is to be expected above all from nuclear resonance tomography investigations.

6.3. Extracerebral Tumours of the Posterior Cranial Fossa (Fig. 11)

These are neurinomas and meningiomas of the cerebellopontine angle region. The CTs in which there were signs of occlusion hydrocephalus are also not considered here.

The cistern alterations in these tumours are characterized by the displacement of the brain stem away from the tumour. In tumours of the medial cerebellopontine angle region, this does not entail a pronounced lateral and smaller sagittal component, which also brings about a slight displacement away from the clivus besides the displacements to the opposite side. Tumours located more laterally do not display this component. Similar to the cerebellar hemisphere

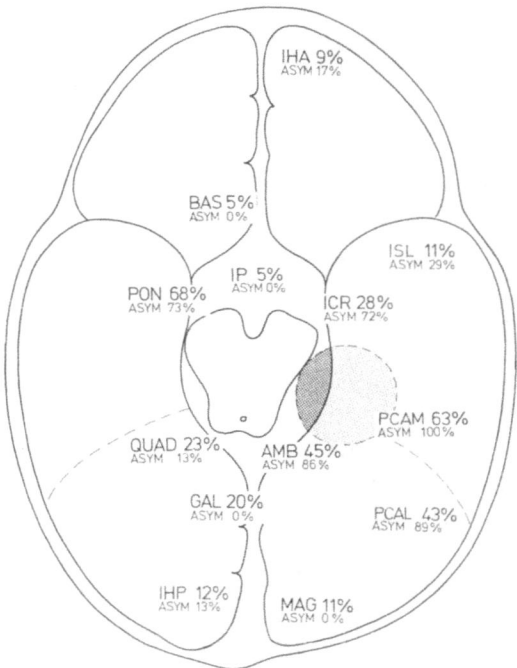

Fig. 11. Fossa posterior: extracerebral tumours

tumours, they tend to lead to compression of the pontine cistern, which may be asymmetrical.

The alterations of the supratentorial spaces remain slight, after the hydrocephalic findings were sorted out. Above all, there is no noteworthy involvement of the basal and intercrural cisterns.

The ambient cistern and the intercrural cistern with their pronounced side differences (86% and 72% respectively) reflect the lateral component of the brain stem displacement. The frequent compression of the quadrigemina and Galen cistern (23% and 20% respectively) and their low asymmetry indicate ascending herniation in simultaneous narrowing of the pontine cistern (55%), with simultaneous dilatation of the pontine cistern a dorsal displacement of the brain stem.

The asymmetrical involvement of the pontocerebellar cistern both in its medial (63%) and its lateral part (43%) results from tamponade by the tumour.

Summary: the tumours of the cerebellopontine angle lead to characteristic alterations of the pontine cistern, which are mainly asymmetrical because of the lateral location of the tumours. The simultaneous displacement of the brain stem away from the clivus causes its dilatation, simultaneously also confinement of the dorsal parts of the perimesencephalic cistern ring. Large tumours can once more lead to a complete cistern blockade and an upward herniation of the cerebellum.

7. Occlusive Hydrocephalus (Figs. 12 and 13)

Above all with regard to the differences to be expected in the behaviour of the perimesencephalic cisterns, the occlusive hydrocephalus elicited by supratentorial tumours (Fig. 12) must be considered separately from those of infratentorial origin (Fig. 13). Common to both is the distinct symmetry of the cisternal alterations, which mostly masks the mass displacement by the underlying tumours.

The side differences due to the tumour can be manifested to a certain degree only at the perimesencephalic cistern ring. This is more pronounced in the infratentorial tumours, in which the ambient cistern is asymmetrical in 38% as compared to 30% in the supratentorial cistern, and 23% in the intercrural cistern (compared to 18%) and 14% in the quadrigeminal cistern (compared to 12%).

In the posterior fossa, the tumours localized there also characteristically affect the cisterns: 43% asymmetry in the pontine cistern, 76% in the lateral cisterns and 82% in the medial pontocerebellar cistern.

In the supratentorial cavity, only the CSF cavities of the interhemispheric fissure can also provide weak pointers with regard to the which side the tumour is present with 16% and 11% side differences even in presence of hydrocephalus.

the interhemispheric fissure clearly stands out in the two groups (90% infratentorial and 69% supratentorial tumours). This indicates a preferential dilatation of the occipital horns of the lateral ventricles, especially since a direct influence of the lateral ventricles does not occur in tumours of the infratentorial cavity, as is in the rule in supratentorial tumours.

Furthermore, the high rate of compression of the basal cisterns is noticeable. This is explained from the protrusion forward of the dilated third ventricle into the basal cistern and interpeduncular cistern. The cistern of Galen and quadrigemina are likewise affected to an unusually large extent, which indicates craniocaudal displacement of the brain stem in the sense of "transtentorial" herniation. The dilatations of these cisterns are more difficult to interpret. In part, they have occurred following a drainage operation, although complete regression of the ventricular dilatation could not be observed. If they are to be observed in the region of the perimesencephalic cisterns even in full hydrocephalus, they can be regarded as signs of asymmetry of the underlying tumour.

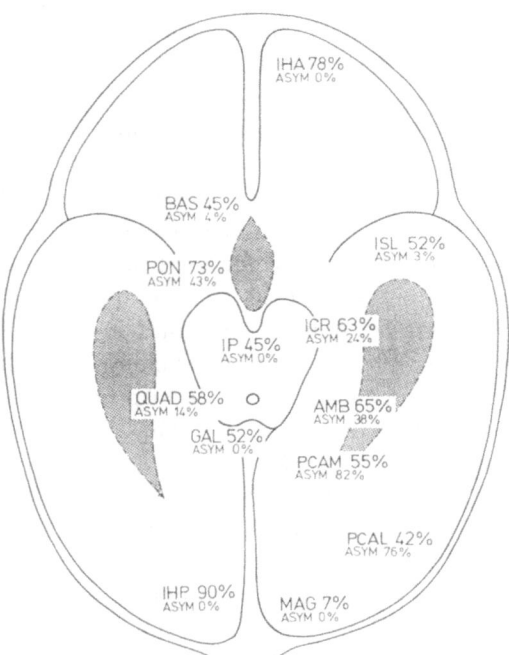

Fig. 13. Occlusive hydrocephalus II

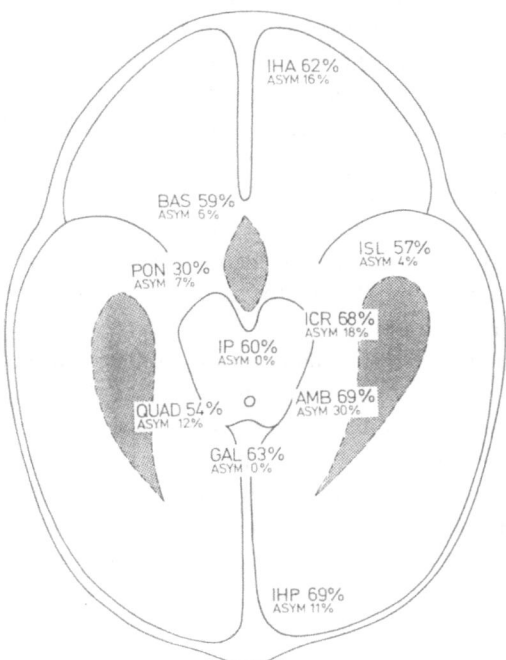

Fig. 12. Occlusive hydrocephalus I

Almost all cisterns are affected by alterations in roughly the same magnitude, but the occipital region of

To summarize, the alterations of the perimesencephalic and basal cisterns are very similar with regard to their symmetry and their percentage distributions in occlusion hydrocephalus irrespective of whether this is of supratentorial or infratentorial genesis. Differences essentially result in the cisterns of the posterior fossa, which also display a greater asymmetry in the infratentorial tumours.

Discussion

There has so far been no systematic description of cistern behaviour in CT in chronic space-occupying intracranial processes. The fundamental knowledge on alterations of cerebral cisterns has essentially been provided by German researchers[14, 19, 21, 22, 23, 24, 27, 28, 29, 30, 31, 32, 33, 36, 37, 38] who investigated the principles of these alterations in the "form" of the brain on the basis of pathological-morphological findings. For understanding the pathogenesis of the most diverse "secondary" alterations in space-occupying processes within the cranial capsule and the clinical consequences resulting from this, these studies are of pre-eminent importance. They only cover the mechanical aspect of the process, but this is often the factor eliciting a series of secondary manifestations which are decisive for the clinical course and are of special interest for treatment. In particular for the neurosurgeon, knowledge of the laws governing the mechanical factors and the reactions they cause is one of the preconditions for rational and effective intervention on his part.

Computer tomography directly shows alterations in form and mass displacement which provide valuable criteria for the evaluation of the situation and the decision regarding therapy and above all surgical measures. The involvement of the cisternal cavities plays a particular role here, in acute as well as in chronic space-occupying processes. Compared to all morphological investigations, it is possible on the basis of computer tomography to study the alterations which occur in space-occupying processes of the most diverse size and location in living patients. Thus, above all findings are available which are characteristic for alterations which are compatible with life and have not in many cases caused the death of the patients as in autopsy investigations. Thus, the alterations in form of the cisternal spaces which are regularly formed by the location and size of the respective tumour can be followed very well in the example of intracranial tumours.

The computer tomographic investigations have primarily revealed that the cisterns react in quite different ways to the space requirement of a tumour. Thus, even in small tumours not only are the CSF spaces in the immediate vicinity compressed, but above all the subarachnoid spaces over the convexity and in the interhemispheric fissure are affected at an early stage. They are thus especially fine indicators for narrowing of the intracranial cavity.

In contrast to this are the cisterns which resist even major deformations, e.g. in large tumours. Thus, the interpeduncular cistern which is surrounded by the powerful fiber masses of the crura cerebri is confined almost exclusively by tumours penetrating directly into it. Another example is the cistern of Galen, which is protected by the eave of the cerebellar tent, and which is compressed exceedingly rarely in supratentorial mass displacements and only with a massive increase in infratentorial volume.

With regard to the tumour location, the rule applies that the mass displacement to the opposite side is the greater the farther away the tumour is from the midline. Thus, tumours of the sphenoid bone, the sellar region or in the vicinity of the third ventricle can cause an confinement of the neighboring cisterns (partly by growing into these spaces), but on the other hand rarely bring about a mass displacement to the side. In consequence of this, they may occasionally reach a considerable size before symptoms occur owing to secondary effects of the mass displacement. The computer tomographic investigations of the perimesencephalic cisterns also show that the suspected "axial displacement" of the brain stem plays practically no role in supratentorial tumours near to the midline. But axial displacement of other structures does occur, for instance cerebellar herniation into the foramen magnum, even if it could not be demonstrated computertomographically in our material from technical reasons.

The observations of the perimesencephalic cisterns are also of diagnostic importance. These in turn reveal even very slight alterations in the position of the brain stem compared to the tentorial notch. Thus, for example small tumours of the cerebellopontine angle region which may occasionally not be subject to direct computer-tomographic detection are revealed by the asymmetry of the ambient cistern which they cause.

Impressive are the massive alterations in occlusive hydrocephalus which affect the majority of the cisterns. As a rule, they mask the effects on the cisterns caused by the tumour itself. Only when the hydrocephalus has been eliminated (e.g. by drainage) do the alterations in the cisterns caused by the tumour become apparent.

The observation that the posterior horns of the lateral ventricles are dilated in occlusive hydrocephalus also appears to be important in contrast to the assumptions based on autopsy investigations[26].

Besides the new knowledge and corrections of

previous assumptions, it is obvious that computer tomographic investigations would also confirm numerous observations already made. Thus the various types of herniation of the uncus gyri hippocampalis can thus be excellently identified. The "champagne cork phenomenon" of Spatz is also imaged in CT as well as the "ascending herniation" of the cerebellum in infratentorial tumours.

A corresponding study of the greater cistern was not carried out in this investigation, since the characteristics of the instrument available did not allow an unequivocal evaluation.

References

Part I

1. Adams DR (1969) The late effects of head injury. Walker AE, Caveness WF, Critchley M (eds). Thomas, Springfield, pp 524–526
2. Alquier M (1905) Deux cas d'hétérotopie du cervelet dans le canal rachidien. Rev Neurol (Paris) 13: 1117–1118
3. Arnold J (1984) Mycelocyste, Transposition von Gewebskeimen und Sympodie. Beitr path Anat (Jena) 16: 1
4. Bannwarth A (1935) Zur Pathologie des Hirntumors. Arch Psychiatr 103: 471
5. Le Beau J (1938) L'oedème du cerveau. Thesis de Paris
6. Cairns H (1952) Disturbances of consciousness with lesions of the brain stem and diencephalon. Brain 75: 109–146
7. Chakeres DW (1983) Brain stem and related structures: normal CT anatomy using direct longitudinal scanning with metrizamide cisternography. Radiology 149 (3): 709–715
8. Chiari H (1986) Über Veränderungen des Kleinhirns, des Pons und der Medulla oblongata infolge congenitaler Hydrocephalie des Großhirns. Denkschr Akad Wiss Wien 63: 71
9. Cushing H (1901) Concerning a definite regulatory mechanism of the vaso-motor centre which controls blood pressure during cerebral compression. Johns Hopkins Hosp Bull 12: 290–292
10. Cushing H (1917) Tumours of the acusticus and the syndrome of the cerebellopontine angle. WB Saunders, Philadelphia
11. Flanningan BD et al (1985) Magnetic resonance imaging of the brain stem: normal structure and basic function anatomy. Radiology 154 (2): 375–383
12. Gehuchten P van (1937) Le mécanisme de la mort dans certains cas de tumeur cérébrale. Encéphale 2: 113–127
13. Han JS et al (1984) Magnetic resonance imaging in the evaluation of the brain stem. Radiology 150 (3): 705–712
14. Hasenjäger Th, Spatz H (1937) Über örtliche Veränderungen der Konfiguration des Gehirns beim Hirndruck. (Zisternenverquellung und Verschiebung über die Medianebene.) Arch Psychiat Nervenkr 107: 193
15. Hawkes RC et al (1983) Craniovertebral junction pathology: assessment by NMR. AHNR 4 (3): 232–233
16. Henschen F (1910) Über die Geschwülste des Kleinhirnbrückenwinkels. Fischer, Jena
17. Jefferson G (1938) The tentorial pressure cone. Arch Neurol Psychiat 40: 857
18. Mawad ME et al (1983) Computed tomography of the brain stem with intrathecal metrizamide. Part I: The normal brain stem. AJR 140 (3): 553–563
19. Meyer A (1920) Herniation of the brain. Arch Neurol Psychiat 4: 387–400
20. Nakagama Y et al (1980) Clinical implications of the CT appearance of the brain stem cisterns in cases of acute and severe head injuries. Prog Comp Tomog 2: 235–241
21. Pia HW (1953) Die Verquellung der Cisterna Basalis und Ambiens im Hirngefäßbild. Acta Neurochir (Wien) 3: 315
22. Pia HW (1955) Die Pathogenese der Gefäßschäden der Occipitallappen bei gesteigertem Hirndruck. Proc Sec Int Congr of Neuropathology. Excerpta medica, Amsterdam, pp 317–319
23. Pia HW (1956) Die Einwirkung der Hirndrucksteigerung auf den Hirnstamm, ihre Klinik und Behandlung. Münch med Wschr 98: 1609–1612
24. Pia HW (1957) Die Schädigungen des Hirnstammes bei den raumfordernden Prozessen des Gehirns. Acta Neurochir (Wien) [Suppl] 4. Springer
25. Planitzer J, Zschenderlein R (1983) Möglichkeiten und Grenzen computertomographischer Hirstamm-Diagnostik. Psychiatr Neurol Med Psychol Beih 29: 205–214
26. Riessner D, Zülch KJ (1940) Über die Formveränderungen des Hirns (Massenverschiebungen, Zisternenverquellungen) bei raumbeengenden Prozessen. Dtsch Z Chir 121: 253
27. Spatz H (1929) Die Bedeutung der "symptomatischen" Hirnschwellung für die Hirntumoren und für andere raumbeengende Prozesse in der Schädelgrube. Arch Psychiat 88: 790
28. Spatz H, Stroescu GJ (1934) Zur Anatomie und Pathologie der äußeren Liquorräume des Gehirns. I–II. Nervenarzt 7: 425–481
29. Spatz H (1939) Pathologische Anatomie der Kreislaufstörungen des Gehirns. Z Neurol 167: 301
30. Tönnis W (1938) Die Entstehung der intrakraniellen Drucksteigerung bei Hirngeschwülsten. Arch klin Chir (Berlin) 193: 669
31. Tönnis W (1939) Zirkulationsstörungen bei krankhaftem Schädelinnendruck. Z Neurol 167: 462 ff
32. Tönnis W, Riessner D, Zülch KJ (1940) Über die Formveränderungen des Hirns bei raumbeengenden Prozessen. Zbl Neurochir 5: 1
33. Tönnis W (1959) Pathophysiologie und Klinik der intrakraniellen Drucksteigerung. Handbuch der Neurochirurgie I, 1. Springer, Berlin Göttingen Heidelberg
34. Vincent C, Thiébaut F, Rappoport F (1930) A propos du cone de pression temporal. Rev neurol (Paris) 2: 116 ff
35. Vincent C, David M, Thiébaut F (1936) Le cone de pression temporal dans les tumeurs des hémisphères cérébraux. Rev neurol (Paris) 65: 536 ff
36. Zülch KJ (1959) Störungen des intrakraniellen Druckes. Die Massenverschiebungen und Formveränderungen des Hirns bei raumfordernden und schrumpfenden Prozessen und ihre Bedeutung für die klinische und röntgenologische Diagnostik. In: Handbuch der Neurochirurgie, I,1. Springer, Berlin Göttingen Heidelberg, pp 208–303
37. Zülch KJ (1961) Über die Entstehung und Lokalisation der Hirninfarkte. Zbl Neurochir 21: 158–178
38. Zülch KJ, Mennel HD, Zimmermann V (1974) Intracranial hypertension. Handbook of clinical neurology, vol 6. North Holland Publ Comp, pp 89–149

Part II

Clinical Findings in Primary and Secondary Brain Stem Lesions

Introduction

Damage to the brain stem is manifested clinically in functional disorders which can be assigned either in isolation or in combination with the following categories:

— disorders of consciousness

— stimulation or deficit manifestations on the part of the cranial nerve nuclei located in the brain stem and the pertinent nerve tracts and connectional systems

— stimulation or deficit manifestations on the part of the autonomic centers

— losses of functions based on lesions to the ascending and descending tract systems passing through the brain stem.

From the combination of lesions affecting the different systems, there are good possibilities of localizing and determining the extent of damage, thus for example in the numerous vascular syndromes of the brain stem with their frequent very circumscribed lesions. These present a varied range of differentiated symptom combinations corresponding to the individual variation of vascular supply.

Diseases which primarily involve the brain stem without affecting other structures may be manifested in a similar way. Tumours, abscesses, granulomas, inflammatory or atrophic processes, spontaneous haemorrhages of various aetiology, even traumatic primary brain stem damage frequently permit a clear topical assignment of the symptoms they cause to their anatomical location.

The differentiation and assignment of the symptoms becomes ever more imprecise in increasing extent of the lesion, the rate of their development being important. Thus rather uniform combinations of symptoms develop which can be assigned to whole regions of the brain stem in damage which is of acute occurrence or very extensive. If such functional disorders involve the entire cross-section of the brain stem, then one can refer to "lesion levels".

This applies above all to damage to the brain stem occurring secondarily in consequence of space-occupying intracranial processes localized outside the brain stem. Although circumscribed morphological alterations occur here, the clinical picture regularly shows signs of functional disorders of transsectional extent. If the process progresses, the brain stem is affected to an ever greater extent and there follow clinical symptom groups characterized by increasing loss of function in a regular time sequence, even if at a different rate.

Independently of the nature of the primary process, the secondarily elicited alterations of the brain stem take over the leading role in the clinical picture of disease and mostly also determine their course and thus the fate of the affected patient[3, 4, 15, 18, 19, 37, 38].

For the prognosis, the reversibility of functional losses suffered is of crucial importance, especially in brain stem lesions[23]. A complete restoration of function allowing inferences with regard to a transient purely functional disorder without persistent morphological alterations is to be observed in secondary involvement of the brain stem only in individual cases. Precondition for this is the only brief duration and an injurious effect on the brain stem which does not go beyond a certain degree. Longer lasting and more extensive damage bring about permanent morphological and functional alterations[3, 4, 16, 17, 18, 30, 36, 43, 44, 45].

On the other hand, the prognosis of irreversible damage to the brain stem is not as unfavourable as was assumed for a long time[9, 11, 21, 23]. This information is primarily due to computer tomography, which reveals numerous substantial brain stem lesions which previously not could be detected in vivo[1, 2, 24, 28, 61, 62, 81]. On the other hand, the possibilities of intensive medicine helped even patients with more extensive lesions to survive in that they enable the critical initial phase with decompensation of the regulatory functions to be bridged over.

Despite the spatial concentration of numerous vital function systems in the brain stem, it has been shown (likewise by means of computer tomography) that even more extensive lesions can be survived. For acute processes, this applies above all when structures associated with intact consciousness are spared. On the other hand, chronic processes can lead to high-grade deformation, displacement, extensive infiltration or marked pressure atrophy of the brain stem, and nevertheless remain low in symptoms. In these patients, the danger of an acute decompensation is great and dangerously restricts its tolerance range, even with regard to surgical operations. Nevertheless, the prognosis of these patients can be more readily appraised with consideration of the underlying process than is the case in patients admitted to the hospital with acute signs of brain stem involvement.

The rather uniform clinical picture of a brain stem lesion (coma, decerebration) in the acute phase makes it difficult to answer the questions as to the extent and possible reversibility of the functional disorders which are important for evaluation of the prognosis. Analysis of clinical findings which may provide particular information on the further course or the expected outcome of the disease therefore gains significance. In other words, what information on the functional state of the brain stem, the course and prognosis of the disease can be inferred from the clinical neurological findings as far as these can be determined for brain stem damage?

McNealy and Plum (1962) assigned certain groups of clinical symptoms for the first time in a systematic way to circumscribed lesion levels of the brain stem. They showed the regular time sequence of symptomatology which results in damage to the brain stem progressing in the rostro-caudal direction (early diencephalic—late diencephalic—midbrain/upper pons —lower pontine/upper medullary).

Numerous other authors assigned clinical symptom groups to damage levels of the brain stem in a similar way, mostly with the intention of attaining of clinical grading with regard to prognostic appraisals[10, 18, 33, 37, 49, 50, 70, 71, 76].

Besides the appraisal of the state of consciousness by capacity for verbal communication, eye opening and motor reactivity[56, 58], functional tests of the nuclear areas and tract systems localized in the brain stem increasingly became the focus of investigations. For this purpose, the brain stem reflexes which can normally be evoked were used[7, 9, 26, 35, 47, 48, 50, 82, 86, 92] as well as the reflexes appearing under pathological conditions[5, 6, 22, 25, 33]. Electromyographic investigations have made important contributions here[5, 6, 22, 25, 31, 32, 34, 51, 52, 53, 63, 77, 84, 85]

Moreover, tests of the function of various tract systems by the investigation of evoked potentials is being used to an increasing extent[12, 39, 40, 41, 55, 56, 57, 58, 59, 60, 74, 89].

Numerous attempts at early prognostic appraisal of brain stem lesions have been made[20, 23, 68, 69, 70, 76, 83, 87, 88, 90, 93], although with very diverse results. In the majority of these studies, the location and the degree of brain stem damage could not be established by computer tomographic investigations and were only documented by autopsy in some cases. Prospective studies with comprehensive documentation of the clinical neurological findings are also rare amongst these studies.

Questions posed

1. What clinical parameters show the closest correlations with the outcome of the disease (i.e. permit prognostic appraisals)?
2. Are there certain constellations of clinical parameters which are associated especially closely with the course or the outcome of the disease?
3. What is the role played by the location and extent of brain stem lesion for the prognosis?
4. From what level of damage is recovery still possible, or is there a point of no return?
5. What prognostic conclusions result from the course of the brain stem symptoms in the first days?
6. From what time are prognostic appraisals possible with adequate safety on the basis of the clinical findings?

Patients and Methods

Between 1980 and 1982, 79 patients with primary and secondary brain stem damage of different aetiology from the Department of Neurosurgery, Justus-Liebig-University of Giessen, Federal Republic of Germany were investigated.

Twenty-nine of these patients had suffered a severe craniocerebral trauma, seven patients with primary involvement of the brain stem on the basis of the clinical, computer tomographic and in some cases pathological anatomical findings. In the remaining cases, there was a secondary involvement of the brain stem on the basis of intracranial space-occupying complications (intracerebral and extracerebral haematomas, oedema).

Twenty-two patients had intracranial tumours (seven supratentorial, 15 infratentorial, of these six deriving from the brain stem).

Twenty-two patients had spontaneous intracranial haemorrhages, 14 in supratentorial location, four with primary involvement of the brain stem, seven infratentorial, of these four primarily in the brain stem.

Three patients with infarcts in the basilar flow area, one patient with acute encephalitis, one patient with acute hypoxic damage and one patient with olivo-pontocerebellar atrophy were likewise included in the study.

In all patients (n = 79), the disease had led to involvement of the brain stem of different degrees. The majority of the patients (n = 57) were admitted with acute disease manifestations. Thirty-four patients died within the first three weeks, eight were transferred to other hospitals in an apallic state. Eleven patients recovered completely, 17 were discharged or transferred with slight and nine with severe neurological deficits.

The results of the total of 509 clinical neurological investigations were recorded on a documentation sheet which registers 1,141 single parameters for investigation.

During the first week of in-patient treatment, the patients were examined daily, whereas in the second week an investigation was carried out every second day and two investigations per week from the beginning of the third week. The very much shorter hospital stay in many of our cases (on average ten days) often only allowed a correspondingly shorter period of observation. All investigations in which pronounced pharmacotherapeutic influences were present (e.g. relaxation in controlled ventilation) could not be considered.

Between one and 22 investigations were carried out per patient. Five hundred and nine documented and evaluated investigations are distributed to 79 patients. In 210 investigations, a two-channel EEG was recorded. In 182 investigations, the intracranial pressure was registered. An autopsy was carried out in 19 patients.

Patients under drugs with an appreciable influence on the neurological findings (sedatives, hypnotics, muscle relaxants) were excluded from the evaluation. Other iatrogenic influences (intubation, diagnostic interventions or surgery) were recorded and used in the evaluation of the investigations.

On recording of the neurological findings, it was established whether injuries or other diseases of the target organs brought about or simulated alterations in the neurological status in addition to the actual disease process (limb loss, fractures, peripheral nerve injuries, pre-existing diseases such as old ischemic infarction, poliomyelitis, diabetes etc.). Especially characterized were patients in whom a complication (e.g. of cardiopulmonary, renal or infectious genesis) determined the course of the disease.

The day of the event (trauma, bleeding, loss of consciousness etc.) or the day of operation served as zero point on the time axis for the clinical course when signs of brain stem involvement only occurred from this point onwards. In chronic processes, the first of the month or the year in which signs of the disease had first occurred served as reference date for the beginning of the illness. As zero point on the time axis, the day of hospital admission was specified in order to obtain a practical time raster.

All data recorded were entered into a survey sheet. From there, they were transferred to marking sheets which can be read by machine from which the data are read onto magnetic tape via punched cards. Further processing of the data was carried out by means of the hospital's own Dietz 621/8 process computer after a large number of plausibility tests and controls had been carried out.

Since mainly counted frequencies were to be worked with, a chi^2 test was appropriate as a statistical test technique, as a rule as a four field test (with and without continuity correction). From the calculated test parameters, the contingency coefficients were determined when comparative appraisals were to be made with regard to the correlation between two characteristics. Application of more far-reaching statistical analyses was dispensed with, since very quickly the case numbers frequently became too small in order to provide statistically substantiated results because of the necessary breakdown of the data material in terms of several parameters.

Results

In the long term, the prognosis of a patient is determined by the nature of the underlying disease and hence only depends to a limited extent on the momentary condition of the patient. The less ominous and the less serious the influence of the disease symptoms on the patients, the lower their prognostic significance, since other factors such as aetiology, location and pathological quality of the disease will then be more important for the long-term prognosis.

However, the more acute and serious the manifestations of the disease, the more distinctly direct connections with the further course of the disease become

apparent from the severity and constellation of clinical symptoms[10, 11], i.e. certain, short ot median term prognostic considerations are primarily supported on the observation of the momentary state of the patient and on the evaluation of any alterations in this state.

In all the patients we investigated (n = 79), the disease had led to varying degrees of involvement of the brain stem. The majority of the patients (n = 57) were admitted with acute signs of disease.

Many measurements and observations could not be made because of the condition of the patient. This explains the different total number of individual parameters. Some findings did not reach adequate reliability because the number of cases which could be evaluated was too small. However, it is to be suspected that these findings are only to be obtained to a reduced extent in any other mixed patient material, so that their significance remains restricted in practice.

1. Individual Parameters and Their Combinations (Tables 1 and 2)

For evaluation of the condition and the associated prospective prognosis of a patient, all available information and thus also the totality of the clinical neurological findings are always to be considered. Nevertheless, it is clear that certain have greater weight than others.

This doubtless applies to the evaluation of the state of consciousness, as practiced for example in the form of the Glasgow coma scale[99]. The significance for the prognosis of the patients' state of consciousness at admission and in the further course is underscored by our results.

The following principle applies: the better the state of consciousness, the higher the reliability of the appraisal with regard to the prognosis. The waking, fully oriented patient had by far the greatest probability of reckoning with a favourable result of treatment (between 93% and 100%) owing to his initial functional state. However, this applies with the restriction that this prognosis related to the period of treatment. The long-term prognosis determined by the nature of the underlying disease is not affected by this prediction.

Conversely, the prognostic value of the state of consciousness becomes less, the poorer this is. Whereas clear consciousness with maintained motor and verbal reaction capacity is linked to intact function of the important parts of the brain stem, disorders of consciousness may be caused by numerous more or less serious alterations. However, these do not always

Table 1. *Examination at admission: correlation of clinical parameters with the outcome (GOS)*

corrected contingency coefficient	parameter	significance (chi-square test with Yates corr.)
0.77	motor reaction	p ≤ 0.01
0.76	light reaction direct	p ≤ 0.01
0.76	eye-opening	p ≤ 0.01
0.75	light reaction indirect	p ≤ 0.01
0.75	spinociliary reflex periph.	p ≤ 0.01
0.73	verbal contact	p ≤ 0.01
0.71	oculocardiac reflex	p ≤ 0.01
0.69	orbicularis oculi reflex	p ≤ 0.01
0.63	eyes open/closed	p ≤ 0.01
0.61	muscular tone arms	p ≤ 0.01
0.58	masseter reflex	p ≤ 0.01
0.55	vestibulo-ocular reflex	p ≤ 0.01
0.50	pupillary width	p ≤ 0.01
0.49	abdominal skin reflexes	p ≤ 0.01
0.48	Babinski phenomenon	p ≤ 0.01
0.45	respiration	p ≤ 0.01
0.43	resp. frequency	p ≤ 0.025
0.36	corneal reflex	n. sig.
0.34	heart rate	n. sig.
0.33	oculomotor paresis	n. sig.
0.29	plantar response	n. sig.
0.26	shape of pupil	n. sig.
0.25	blood pressure	n. sig.
0.21	body temperature	n. sig.
0.15	tendon reflexes	n. sig.

necessarily entail a poor prognosis. Of our patients admitted in a somnolent or comatose state, 70% to 77% had a poor prognosis. The remaining patients survived with different degrees of damage. Especially in patients with clouded consciousness or in comatose patients, it is more urgent to appraise the situation and to arrive at prognostic assessments. This gives rise to the need for more precise information.

By mutual combination of the parameters characterizing the state of consciousness, an appreciable enhancement of the reliability can be attained, even if the contingency coefficient (0.88 to 0.89) indicates that a greater measure of agreement is to be expected. This owing above all to the fact that all three parameters are only different aspects of the same function, namely the state of consciousness. Only introduction of a parameter independent of the state of consciousness also provides higher correlations with the prognosis for comatose patients: 85% and 88% of the patients who did not present any light reactions of the pupils in addition to the lack of motor, verbal and eye opening reaction had a poor prognosis.

On the other hand, prognostic reliability of this level is almost reached even with evaluation of the *light reaction* alone. This was one of the parameters displaying the greatest agreement with the result of treatment (contingency coefficient = 0.76 in the direct light reaction, contingency coefficient = 0.75 in the indirect light reaction). Eighty-seven percent of the patients with attenuated or distinguished pupil reaction belonged to the group with a poorer prognosis. The close correlation between the lack of pupillary reflexes and a poor outcome showed the greatest prognostic reliability both in the single parameters and in the combination of findings: patients in whom both the direct and indirect light reaction was lacking had a poor prognosis in 91%. With regard to the favourable evaluation in maintenance of both reflexes, this combination remained behind the prognostic reliability of the state of consciousness with a correlation of 85%.

The *spinociliary reflex* evoked peripherally surprisingly appears in the group of parameters showing the greatest agreement with the result of treatment. This may be explained by the fact that this is a reflex which tends to be stable in brain stem lesions, and which also remains stable even in extensive or advanced damage to the brain stem. Its stability is based on a reflex arch which is still intact even in high cervical spinal damage: afferents via dorsal roots to synapses in the upper thoracic spinal cord, ascending fibers via upper thoracic roots to the stellate ganglion, and from there after renewed synapsing running with the internal carotid artery, further connecting the ophthalmic nerve to the ciliary ganglion, and from there to the pupillary dilator muscle[85]. Although for this reason it cannot be directly used to evaluate brain stem lesions, the peripheral spinociliary reflex nevertheless provides good prognostic pointers. Thus 83% of the patients with maintained reflex with GOS five to three survived, whereas 85% of the patients in whom it could no longer be evoked belonged to the group with GOS two or one.

The interpretation of the results is more difficult in *oculocardiac reflex*. In terms of its physiology, it should belong to the stable reflexes in brain stem lesions (similar to the spinociliary reflex) and should still be preserved in damage in the pontine region. Thus a greater prognostic reliability would be expected in the case of its extinction, whereas there should be a lower correlation with the prognosis in its absence. It affords a good contingency of 0.73, but its reliability with 94% favourable prognosis is higher when its function is preserved than when it is lost, being associated with a poor outcome in only 73% in the latter case. It is possible that the reflex has more complex synaptic links than was to be assumed on the basis of investigations so far.

Another effect appears to be have caused the high reliability of the *orbicularis oculi reflex*. As a typical exteroreceptive reflex, its function is linked to the integrity of several structures localized inside and outside the brain stem[27, 31, 34, 42, 63, 65, 66]. It is hence one of the first reflexes to be attenuated or abolished for example in secondary brain stem damage, frequently already in the diencephalic stage. However, if it is preserved, it indicates in a significant way a more favourable prognosis: 82% as compared to 32% in its absence.

Similar situations are also to be postulated for the *central spinociliary reflex, i.e.* that evoked in the area innervated by the trigeminal nerve. Its function depending on the intactness of the structures mediating the reflex in the brain stem (sensory afferences via the trigeminal nerve to the nuclear area in the diencephalon = posterior and lateral hypothalamus, descending fibers in the prerubral field, dorsal and rostral connection to the red nucleus, via the lateral mesencephalic tegmentum, pons, medulla ascending to synapses in the upper thoracic spinal cord, from there further via the same efferents as the peripherally evoked reflex[6]) is frequently no longer maintained even in less pronounced involvement of the brain stem owing ot this complex reflex arch. It did not show any prognostic significance in our patients.

The second group of clinical parameters which possess a low but still significant correlation with the prognosis is headed by the *spontaneous position of the eyelids*. As a yardstick for the state of activation of the reticular formation, open eyes indicate the function of major parts of the reticular formation. If the patient's eyes are closed in the investigation, this does not automatically indicate a lesion of the reticular formation. Only the evaluation of eye opening in response to waking stimuli permits an appraisal.

The evocability of the *deep tendon reflexes and pareses of the limbs* did not show any significant correlation with the prognosis. Both parameters are subject to alterations which may be caused by processes outside the brain stem.

On the other hand, *muscle tonus* shows a correlation with the outcome of the disease. Precondition for this is that the various pathological tonus forms are not considered singly, but that they are grouped together.

Furthermore, it is to be considered that these are findings at the investigation on admission, i.e. disorders of muscular tonus which have occurred acutely, and which are a manifestation of brain stem involvement in the majority of cases.

In contrast to the other deep tendon reflexes, the *masseter reflex* has a significant correlation with the outcome groups. It once more has features of a "stable" reflex absence of which provides more prognostic information than preservation of its function.

Conversely, the *vestibulo-ocular reflex* shows the characteristics of a "complex" reflex, the function of which is associated with the intactness of numerous structures. Since it is already lost even in brain stem lesions of smaller extent, the prognostically favourable assessment in its maintained function has the higher correlation with the outcome.

Pupillary width is significantly associated with the outcome of the disease. However, with regard to the extent of its coherence it is already in the borderline region. Several factors contribute to this limited prognostic reliability. The evaluation of pupillary width is already subject to the subjective influences of the different investigators in comatose patients under the usual clinical conditions. The different pupil width (constricted = diencephalic or pontine stage; moderately dilated = mesencephalic stage or normal finding; highly dilated = mesencephalic or bulbar stage) in the different lesion levels is thus assigned to different prognostic expectations. Peripheral lesions of the oculomotor nerve or of the eye as well as pharmacotherapeutic influences on pupillary width must likewise be able to be ruled out beforehand. In the final analysis, pupillary width has great prognostic reliability only in connection with the evaluation of the light reaction.

The prognostic rating of the *abdominal skin reflex* is likewise in the borderline range. As exteroceptive and thus "complex" reflex, they do not have very great prognostic reliability in their absence, whereas their presence is a clear indication of a favourable course.

As to be expected, the *Babinski phenomenon* only provides a few prognostic indications in its absence. However, if it can be evoked in the acute stage, it shows a just still significant correlation with the prognostically poorer group.

In the *evaluation of the autonomic parameters*, the number of investigations which can be utilized only affords a slight possibility of statistical analysis. In the comparison of adequate spontaneous breathing and all other forms of *disturbances in breathing* there are still significant group differences, as well as in the breathing

rate, when the values deviating from the normal are taken together. The same applies to the frequency of breathing when all values deviating from the normal are summarized. Since this summary lumps together very different conditions, the boundary of the reasonableness of statistical analysis is probably reached. Here, there is no doubt that consideration of the individual case is to be preferred, especially since other possible influences on autonomic parameters must be ruled out in the individual case.

Some of the clinical parameters showed surprisingly low diagnostic reliability. Thus for example the *corneal reflex*, absence or pathological values of which showed too great a scatter in the two outcome groups in order to be prognostically utilizable. *Pathological reflexes* such as the corneomandibular or the palmomental reflex were manifested too rarely in order to enable appraisals. The *pupil form*, the pathological variants of which were to be observed exclusively in patients with poor prognosis did not reach significance owing to its small number (n = 10).

Losses of cranial nerve function were likewise of low prognostic value. Since most of the cranial nerves can only be detected in cooperative patients, it was not possible to make a comparison with patients having clouded consciousness. The same applies correspondingly to all findings the recording of which requires collaboration by the patient, e.g. sensitivity or coordinative actions, disturbances of higher cortical functions.

Among the *combinations of findings* which displayed a good correlation with the outcome of the disease (contingency coefficient = 0.7) and an adequate number of characteristics (n = 45), the parameters which already display the best coherence as single findings are found exclusively.

Special attention is to be paid to the *combination of direct light reaction and orbicularis oculi reflex,* of which the high contingency of 0.82 shows a high prognostic value. Its prognostic reliability is high both in patients in whom both reflexes are preserved and in patients in whom the two reflexes are abolished. The light reaction with its reflex-mediating structures in the upper midbrain, and the orbicularis oculi reflex with its reflex arc on the pontine or even on the supranuclear level indicate the function of important parts of the brain stem when the two are intact. Conversely, a lesion which causes both to disappear will be more extensive than a damage which only affects one of the two reflexes.

In connection with the motor reaction capacity, the

Table 2. *Examination at admission: correlation of combined clinical parameters with the outcome (GOS)*

number of patients with both parms.	corrected contingency coefficient	combination of parameters	significance (chi-square test with Yates corr.)
50	0.89	verbal contact/light reaction dir.	p ≤ 0.01
47	0.88	motor reaction / l.r. direct	p ≤ 0.01
50	0.88	eye-opening / l.r. direct	p ≤ 0.01
51	0.82	l.r. dir./orbicularis oculi reflex	p ≤ 0.01
56	0.82	l.r. direct / l.r. indirect	p ≤ 0.01
48	0.81	orb. oculi r. / eye-opening	p ≤ 0.01
46	0.81	mot. reaction / orb. oculi r.	p ≤ 0.01
61	0.79	mot. reaction / verb. contact	p ≤ 0.01
62	0.78	mot. reaction / eye-opening	p ≤ 0.01
66	0.74	eye-opening / verb. contact	p ≤ 0.01

orbicularis oculi reflex proves to be a good parameter. It cannot increase the already high diagnostic reliability of the orientated motor reaction in patients with a poor or abolished movement reaction. However, its absence entails the raised probability of an unfavourable prognosis (86%). The orbicularis oculi reflex reaches an almost equally high value in connection with reactive opening of the eyes and verbal contact capacity. It thus proved to be the brain stem reflex with the highest prognostic value after the pupil reaction.

2. Progress Investigations (Tables 3–9)

In our investigations, it has been shown that individual clinical findings are not sufficient to describe progress. Although they may be informative in the individual case, they do not lead to results which are comparable from patient to patient. In order to obtain comparable groups of patients, it was necessary to group together patients with similar degrees of damage to the brain stem, if possible irrespective of their different aetiology.

For description of the functional situation of the brain stem, a classification could be readily applied which is given by the often observed spontaneous course of a progressive secondary brain stem damage in raised supratentorial pressure. With increasing damage progressing in the craniocaudal direction, characteristic clinical symptoms occur which are the manifestation of increasing loss of function. On the basis of these symptoms or symptom complexes, the focus of the lesion or even the "lesion plane" present at the time of the investigation can be readily localized and described. Several authors have utilized this, above all for prognostic reflections in acute craniocerebral injuries[10, 18, 33, 37, 49, 50, 64, 70, 71, 76].

Besides patients with acute secondary lesions of the brain stem, there were also numerous patients with acute or chronic primary lesions of the brain stem. The classification into "lesion levels" could not provide a comprehensive description of the functional state of the brain stem in all cases. For this reason, it also appears not to be reasonable in all cases to refer to "lesion levels", if anything one can speak of "lesion foci". Moreover, some of the clinical symptoms proved to be capable of regression, so that in each case not a structural, but time and again also a reversible functional damage was to be assumed. Hence, the designation "functional stages" is appropriate. Our definition of those "functional levels" is given on Table 3–Table 9.

Among the stages we defined, the "cortical" stage 1 possesses the least prognostic relevance. It subsumes all patients with intact consciousness without taking into account the different locations and severities of lesions of processes. Among these, all patients with chronic condition without any appreciable alteration of their state during the period of observation (e.g. brain stem tumours) are to be found. The prognosis of these patients is crucially determined by the nature and progression of the underlying disease and in the final analysis cannot be inferred from the symptoms at the time of in-patient treatment. On the other hand,

Table 3. *Cortico-subcortical stage (stage 1)*

LEVEL OF LESION	CONSCIOUSNESS	MOTOR SYSTEM	RESPIRATION, AUTONOMIC S.	EYE MOVEMENTS PUPILLARY SIZE	REFLEXES	
CORTICAL OR SUBCORTICAL	NORMAL OR SOMNOLENT	NORMAL OR APPROPRIATE TO PAIN STIMULUS	NORMAL	NORMAL OR ROVING EYE-MOVEMENTS PUPILS NORMAL	ORBIC. OCULI	+
					PUPILLARY	+
					CORNEAL	+
					MASSETER	+
					CILIO-SPINAL	+
					OCULOCEPHALIC	
					vertical	+
					horizontal	+
					OCULOVESTIBULAR	+
					OCULOCARDIAL	+
					PALMOMENTAL	-/+
					CORNEOMANDIBULAR	-
					BABINSKI PHENOM.	-

Table 4. *Diencephalic stage (stage 2)*

LEVEL OF LESION	CONSCIOUSNESS	MOTOR SYSTEM	RESPIRATION, AUTONOMIC S.	EYE-MOVEMENTS	REFLEXES	
DIENCEPHALIC upper/early	SOMNOLENT, RESTLESS OR	APPROPRIATE TO NOXIOUS STIMULI	NORMAL OR	ROVING EYE-MOVEMENTS	ORBIC. OCULI	-
					PUPILLARY	(+)
					CORNEAL	+
lower/late	COMA	MUSCULAR TONE RAISED, begin- ning FLEXION RIGIDITY	CHEYNE-STOKES RESPIRATION	possible DOWNWARD GAZE SMALL PUPILS	MASSETER	+
					CILIO-SPINAL	+
					OCULOCEPHALIC	
					vertical	(+)
					horizontal	+/++
					OCULOVESTIBULAR	+/++
					OCULOCARDIAL	+
					PALMOMENTAL	-/+
					CORNEOMANDIBULAR	-
					BABINSKI PHENOM.	+

Table 5. *Diencephalo-mesencephalic stage (stage 2-3)*

LEVEL OF LESION	CONSCIOUSNESS	MOTOR SYSTEM	RESPIRATION, VEGETATIVE S.	EYE MOVEMENTS PUPILLARY SIZE	REFLEXES	
DIENCEPHALIC- MESENCEPHALIC upper/early	SOPOR OR COMA	MUSC.TONE,REFLE- XES INCREASED FLEXOR RESPONSE (pref. arms)	CHEYNE-STOKES OR	ROVING EYE-MOVEMENTS possible	ORBIC. OCULI	-
					PUPILLARY	(+)
					CORNEAL	+
lower/late	COMA	EXTENSOR RESP.= DECERBRATE RIGI- DITY (pref. legs)	HYPERVENTILATION BP INCREASED TACHYCARDIA HYPERTHERMIA	GAZE DEVIATION DOWNWARD PUPILS MIDDLE SIZE, INTERNUCLEAR GAZE PALSY	MASSETER	+/++
					CILIO-SPINAL	-
					OCULOCEPHALIC	
					vertical	-
					horizontal	(+)
					OCULOVESTIBULAR	(+)
					OCULOCARDIAL	+
					PALMOMENTAL	-
					CORNEOMANDIBULAR	+
					BABINSKI PHENOM.	+

however, the "cortical" stage also comprises patients who regained consciousness in the course of an acute illness.

From these different preconditions, there is nevertheless no disadvantage for our problem, since the classification into functional stages did not have the objective of describing the location of the process for diagnosing the nature of the condition. The stages served exclusively for a statistically measurable progress description, and the prognostic reflections were related exclusively to the clinical state of the patient at the time of investigation. This also means that as in the

Table 6. *Mesencephalic stage (stage 3)*

LEVEL OF LESION	CONSCIOUSNESS	MOTOR SYSTEM	RESPIRATION, AUTONOMIC S.	EYE MOVEMENTS PUPILLARY SIZE	REFLEXES	
MESENCEPHALIC	COMA	MUSC.TONE,REFLE-XES INCREASED DECEREBRATE RIGIDITY	HYPERVENTILATION BP INCREASED TACHYCARDIA HYPERTHERMIA	NO EYE-MOVEMENTS PUPILS of MIDDLE SIZE	ORBIC. OCULI	−
					PUPILLARY R.	−
					CORNEAL	(+)
					MASSETER	++
					CILIO-SPINAL	−
					OCULOCEPHALIC	
					vertical	−
					horizontal	(+)
					OCULOVESTIBULAR	(+)
					OCULOCARDIAL	+
					PALMOMENTAL	−
					CORNEOMANDIBULAR	+
					BABINSKI PHENOM.	+

Table 7. *Mesencephalo-pontine stage (stage 4)*

LEVEL OF LESION	CONSCIOUSNESS	MOTOR SYSTEM	RESPIRATION, AUTONOMIC S.	EYE MOVEMENTS PUPILLARY SIZE	REFLEXES	
MESENCEPHALO-PONTINE	COMA	MUSC.TONE,REFLE-XES INCREASED DECEREBRATE RIGIDITY	HYPERVENTILATION OR "APNEUSIS" BP INCREASED TACHYCARDIA HYPERTHERMIA	NO EYE-MOVEMENTS PUPILS SMALL "PINPOINT"	ORBIC. OCULI	−
					PUPILLARY	−
					CORNEAL	−/+
					MASSETER	++
					CILIO-SPINAL	−
					OCULOCEPHALIC	
					vertical	−
					horizontal	−/+
					OCULO-VESTIBULAR	−/+
					OCULOCARDIAL	+
					PALMOMENTAL	−
					CORNEOMANDIBULAR	−/+
					BABINSKI PHENOM.	+

Table 8. *Ponto-medullary stage (stage 5)*

LEVEL OF LESION	CONSCIOUSNESS	MOTOR SYSTEM	RESPIRATION, AUTONOMIC S.	EYE MOVEMENTS PUPILLARY SIZE	REFLEXES	
PONTO-MEDULLARY	COMA	MUSC.TONE FLACCID NO REFLEXES NO MOTIONS possible FLEXOR RESPONSE	"MACHINE" TYPE OR PERIODIC OR ATAXIC RESPIRATION HYPOTHERMIA BP UNIFORM HR UNIFORM	NO EYE-MOVEMENTS PUPILS WIDE	ORBIC. OCULI	−
					PUPILLARY	−
					CORNEAL	−
					MASSETER	−
					CILIOSPINAL	−
					OCULOCEPHALIC	
					vertical	−
					horizontal	−
					OCULOVESTIBULAR	−
					OCULOCARDIAL	−/+
					PALMOMENTAL	−
					CORNEOMANDIBULAR	−
					BABINSKI PHENOM.	−

evaluation of the prognostic value of individual clinical findings, the long-term prognosis which results from the nature of the underlying disease had to be left out of consideration.

With regard to individual clinical parameters, it was important also to record the findings which occurred without impairing the state of consciousness and without acutely threatening the life of the patient. The chronic processes provide so to speak the model of isolated deficits of individual function and thus permit appraisal of their individual significance. Such an appraisal related to individual function is not possible

Table 9. *Medullary stage ("brain death") (stage 6)*

LEVEL OF LESION	CONSCIOUSNESS	MOTOR SYSTEM	RESPIRATION, VEGETATIVE S.	EYE MOVEMENTS PUPILLARY SIZE	REFLEXES	
MEDULLARY	COMA	MUSC.TONE FLACCID	IRREGULAR,ATAXIC	NO EYE-	ORBIC. OCULI	-
		NO REFLEXES	RESP. ARREST	MOVEMENTS	PUPILLARY	-
		NO MOVEMENTS	POIKILOTHERMIA	PUPILS WIDE	CORNEAL	-
			BP UNIFORM		MASSETER	-
			HR UNIFORM		CILIO-SPINAL	-
					OCULOCEPHALIC	
					vertical	-
					horizontal	-
					OCULOVESTIBULAR	-
					OCULOCARDIAL	-
					PALMOMENTAL	-
					CORNEOMANDIBULAR	-
					BABINSKI PHENOM.	-

in patients with other stages, since in these patients the loss of a function is only an indicator of more extensive damage. In other words: an isolated loss (for example of a single brain stem reflex) does not as a rule has any meaning for the prognosis of a patient. On the other hand, if its loss is a symptom of a more extensive functional disorder, it gains increased significance.

The description of the condition of a patient by functional stages has proved to be useful both for description of the course and for prognostic appraisals. They already divide the patients into groups with an unequivocal prognostic classification at the examination on admission:

Patients who were in stages 5 and 6 on admission have not survived. Patients in stage 4 have either not survived or only survived in an apallic state. On the other hand, all patients with stage 1 on admission have survived. Only in stages 2 and 3 was a final appraisal not possible from the condition on admission alone.

However, in these patients the observation of the course rapidly provided further information. Thus all patients with admission stage 2 had a good prognosis when they had not deteriorated up to the fourth day after admission. None of them died or became apallic. However, the patients who had arrived in stage 2 but who were to be accorded a poorer stage up to the fourth day owing to increasing functional deficits died or became apallic. In the patients with stage 3 on admission, the fourth day after admission likewise provided a clear separation. Those who had improved into stage 1 or 2 up to that time had a favourable prognosis. Only one of these patients was apallic on transfer. The others survived with mainly slight deficits. Those patients whose conditions had deteriorated up to the fourth day died. A "constant course" remaining in

stage 3 entails a poor prognosis for these patients. Apart from one, they died or remained apallic.

In our patients, 80% of the progress alterations had occurred up to the fourth day. All patients in whom an alteration occurred at a later time are in the group of "constant course", since there had been no alteration in their stage up to the fourth day. It becomes evident here that precisely in these patients the prognosis and thus the direction of the later alteration depends on the initial stage. Those who had remained in stage 2 up to the fourth day all had a good prognosis, i.e. the alteration occurring at a later time was an improvement. The patients who had remained in stage 3 or in a poorer stage up to the fourth day all had a poor prognosis with the exception of one patient, i.e. the alteration which occurred later was a deterioration.

Thus, with knowledge of the functional stage, a prognosis can already be made with high accuracy already on the fourth day in almost all of these cases.

3. Summary

The evaluation of the clinical neurological findings in 509 investigations on 79 patients with primary and secondary brain stem lesions has shown that certain neurological findings are particularly able to provide prognostic indications.

Besides the evaluation of the state of consciousness, these are primarily the pupil light reaction, the peripherally evoked spinociliary reflex, the orbicularis oculi reflex, the muscular tonus and the masseter reflex. All patients who died later or remained in an apallic state had normal findings only in a maximum of four out of these eight parameters on admission to the clinic. On the other hand, about 80% of the patients who have survived and who have left the hospital with no, slight or severe neurological deficits, showed normal findings in more than half of these parameters.

In addition, the definition of the lesion focus in the brain stem likewise already provides clear prognostic indication at the examination on admission.

Patients with high-grade functional losses in the oral regions of the brain stem above the pontine level all had a poor prognosis. Patients with partial functional disorder in these regions or in the regions situated further orally had a good prognosis as long as they had improved at least up to the diencephalic functional stage within the first four days. The patients who had deteriorated or who had remained in an unchanged mesencephalic stage within the first four days all had a poor prognosis.

It remains to be pointed out that all the clinical neurological findings together must still be considered in order to localize and diagnose the nature of the brain stem processes. The possible interaction of complications is likewise also to be considered in evaluating the progress. In view of these considerations, the criteria mentioned possess a high degree of prognostic reliability.

Discussion

In the second part of the present paper, patients with different damage in the region of the brain stem resulting from cerebral mass displacement were investigated. By means of electronic data processing, the results were determined from a large number of clinical data which have the highest prognostic reliability with regard to the prognosis and evaluation of the course.

The basic idea of the study is not only to record the abundant clinical data made available in the course of a patient's hospital treatment, but also to make it accessible to rational statistical analysis. At a time at which the improved techniques allow recording and storage of a large number of results from a patient day by day, the need to process the often vast amount of data which leads to utilizable results without simultaneous false appraisals and losses of information concomitant with the data reduction involved in this is growing. For this purpose, on the one hand weighting of the results with regard to the desired information is necessary. On the other hand, the individual results must be incorporated in an overall concept which describes the respective state of the patient investigated in order in this way to attain greater stability of the information obtained and to render improbable a

wrong appraisal owing to erroneous individual results.

The weighting of individual clinical parameters or their combination gives rise to some methodological problems. The basic condition is above all a sufficient number of investigations, since statistical significance is rapidly lost with a decreasing number. It is hence necessary to subsume patients with brain stem damage of different aetiology and thus to accept an inhomogeneity of the underlying clinical pictures. This applies similarly to other factors such as age, sex, concomitant diseases etc. which are important for an individual appraisal, but which are always a possible source of error in statistical analyses. It is clear that patients whose disease course was mainly determined by factors which are not attributable to the brain stem damage were not considered from the start.

The result obtained is remarkable in two respects. On the one hand, it confirms the significance of the investigation of "motor reaction", the "light reaction", the "eye opening" and "verbal contact" which was already substantiated empirically. This can be rated as a cross-check of the reliability of the statistical method. On the other hand, clinical findings to which only slight significance was attached in routine clinical work appear to be important: the "spinociliary reflex" and the "oculocardiac reflex", which are still superior to the "orbicularis oculi reflex" with regard to their diagnostic reliability. In the recapitulation of the physiological fundamentals, this at first glance surprising result is rendered understandable. In my opinion, it also shows that the statistical processing of clinical data can sometimes contribute to elucidating interrelationships which are not discernible at first glance.

The problem of statistical analysis of the disease course on the basis of the clinical data is more difficult to solve. The large multiplicity of alterations of individual findings which moreover have only a rough correlation with the disease course makes it impossible to describe the cause and the basis of individual clinical parameters. Only the description of the momentary state of the patient by a group of parameters which may also be variable in narrow limits leads to results which can be utilized. With the introduction of "functional stages", a state description has been obtained in which any patient with a brain stem damage can be classified at the respective time of investigation on the basis of his clinical findings. The progress then results from the sequence of functional stages.

The focal point of this part of the present study is doubtless the development of methodology. The results obtained confirm fundamentally the experience obtained in the treatment of numerous patients with brain stem lesions. For example, the poor prognosis of patients with pontine or pontobulbar lesion level or the more favourable prognosis in damage to the brain stem which is only partial or located farther orally is confirmed.

A new insight resulting from the statistical progress investigation is that the crucial change in the course decisive for the outcome of the disease had occurred up to the fourth day after beginning of the disease in 80% of the patients investigated. The prognosis of patients whose course had not altered up to that time depended almost exclusively on the level of damage at the time of admission.

The major result of the investigation is that the method developed allows continuous evaluation by means of electronic data processing of the clinical data recorded and stored in such a way that useful appraisals on the prognosis and course are possible in patients with brain stem lesions.

References

Part II

1. Agnoli AL, Cristante L, Busse O, Feistner H (1980) Brain herniation by cranial computer tomography: clinical radiological correlations. In: Anatomy-physiology in CT. Kugler Medical Publications, Amstelveen
2. Agnoli AL, Laun A, Busse O, Schoch R (1980) Computerized tomography in brain stem haemorrhage. Diagnostic and prognostic aspects. Lecture: IXth Congress of European Society of Neuroradiology, Brussels
3. Amphoux M, Sevin A (1975) Traumatisms cérébraux et focalisations intracraniennes. Agressologie 16, A: 47–53
4. Ansari K (1974) Les hémorrhagies secondaires du tronc cérébral. Arch Suiss de Neurol Neurochir Psychiat 114, fasc 1, 1–28
5. Ansink BJJ (1962) Physiological and clinical investigations into four brain stem reflexes. Neurology (Minneap) 12: 320
6. Arieff AJ, Pyzik SW (1953) The ciliospinal reflex in injuries of the spinal cord in man. Arch Neurol Psychiat (Chic) 70: 621
7. Auer LM et al (1980) Predicting letal outcome after severe head injury—a computer-assisted analysis of neurological symptoms and laboratory values. Acta Neurochir (Wien) 52: 225–238
8. Auer LM et al (1980) Relevance of CT-scan for the level of ICP in patients with severe head injury. In: Shulman K et al (eds) Intracranial pressure IV. Springer, Berlin Heidelberg New York, pp 45–47
9. Babić B et al (1979) Prognostic factors in acute head injuries—brain stem contusion during the first week. Acta Neurochir (Wien) 1 [Suppl] 28: 153–157
10. Barge M, Ohanessian J, Baum L, Benabid AL, Chirossel JP (1977) Valeur diagnostique et pronostique des réflexes du tronc cérébral dans les comas post-traumatiques graves. Neurochirurgie 23: 227–238
11. Becker DP et al (1977) The outcome from severe head injury with early diagnosis and intensive management. J Neurosurg 47: 491–502
12. Benezech J et al (1979) Apport des potentiels évoqués du tronc cérébral (PET) pour l'étude clinique et le pronostic des comas traumatiques. Corrélations avec les réflexes du tronc cérébral. Neurochirurgie 25 (4): 219–223
13. Berger MS et al (1985) Outcome from severe head injury in children and adolecents. J Neurol 62: 194–199
14. Bogousslavsky J, Meienberg O (1987) Eye-movement disorders in brain stem and cerebellar stroke. Arch Neurol 44 (2): 141–148
15. Brendler SJ, Selverstone B (1970) Recovery from decerebration. Brain 93: 381–392
16. Bricolo A (1976) Prolonged post-traumatic coma. In: Vinken PJ, Bruyn GW(eds) Handbook of clinical neurology. North Holland Publishing Comp, Amsterdam, Oxford, vol 24, p 699
17. Bricolo A, Gentilomo A, Rosadini G, Rossi GF (1968) Long-lasting post-traumatic unconsciousness. Acta Neurol Scand 44: 512–532
18. Bricolo A, Battistini N, Bergamini L et al (1975) A proposal for the clinical classification of acute coma due to organic cerebral lesions. J Neurosurg Sci 19: 113
19. Bricolo A, Dolce G (1975) Evoluzioni cliniche del coma post-traumatico grave. Min Neurochir 13: 61
20. Bruce DA, Schut L, Bruno LA, Wood JH, Sutton LN (1978) Outcome following severe head injuries in children. J Neurosurg 48: 679–688
21. Bryan R, Weisberg L (1982) Prolonged survival with good functional recovery in 3 patients with computer tomographic evidence of brain stem haemorrhage. Comput Radiol 6 (1): 43–48
22. Buonaguidi R, Rossi B, Sartucci F, Ravelli V (1979) Blink reflexes in severe traumatic coma. J Neurol Neurosurg Psychiat 42: 470–474
23. Burns J et al (1980) Recovery following brain stem haemorrhage. Ann Neurol 7 (2): 183–184
24. Busse O, Agnoli A, Schoch P, Laun A (1981) Primary brain stem haematomas of vascular aethiology: clinical and CT-correlations. 12th World Congress of Neurology, Kyoto, Japan
25. Byuke O (1980) Facial reflex examination. A clinical and neurophysiological study on acustic tumours and brain displacement at the tentorial notch. Acta Neurol Scand [Suppl] 62 (76): 1–127
26. Carlsson C, Essen C von, Lörfren J (1968) Factors affecting the clinical course of patients with severe head injuries. Part 1: Influence of biological factors. Part 2: Significance of post-traumatic coma. J Neurosurg 29: 242
27. Clay SA, Ramseyer JC (1977) The orbicularis oculi reflex: pathologic studies in childhood. Neurology 27: 892–895
28. Cooper PR et al (1979) Traumatically induced brain stem haemorrhage and the computerized tomographic scan: clinical, pathological and experimental observations. Neurosurg 4 (2): 115–124
29. Le Coz P et al (1986) Aspects cliniques et evolutifs des haematomas circonscrits du tronc cérébral. Apport du scanner X. Rev Neurol (Paris) 142 (1): 52–60
30. Crompton MR (1966) Prolonged coma after head injury. The Lancet 938–940
31. Dengler R, Struppler A (1981) Beurteilung der Lokalisation und Ausdehnung von Hirnstammaffektionen mit Hilfe des Orbicularis-oculi-Reflexes. EEG EMG 12 (1): 50–55

32. Denny-Brown D (1962) The midbrain and motor integration. Proc Roy Soc Med 55: 527

33. Fisher CM (1969) The neurological examination of the comatouse patient. Acta Neurol Scand [Suppl] 36: 45

34. Fisher MA, Shahani BT (1979) Assessing segmental excitability after acute rostral lesions: II. The blink reflex. Neurology (Minneap) 29 (1): 45–50

35. Gennarelli TA *et al* (1982) Influence of the type of intracranial lesion on outcome from severe head injury. J Neurosurg 56: 26–32

36. Gerstenbrand F (1967) Das traumatische appallische Syndrom. Springer, Wien New York

37. Gerstenbrand F, Lücking CH (1970) Die akuten traumatischen Hirnstammschäden. Arch Psychiatr Nervenkr 213: 264–281

38. Gerstenbrand F (1970) Klinik und Therapie der akuten traumatischen Hirnstammschäden. Z ges Neurologie Psychiat 197: 105

39. Greenberg RP *et al* (1977) Evaluation of brain function in severe human trauma with multimodality evoked potentials. Part 1: Evoked brain injury potentials, methods and analysis. Part 2: Localisation of brain dysfunction and correlation with post traumatic neurological conditions. J Neurosurg 47 (2) 150–177

40. Greenberg RP *et al* (1979) Clinical findings associated with brain stem dysfunction: an electrophysiological study in severe human head trauma. In: Popp AJ *et al* (eds) Neural trauma. Raven Press, New York, pp 229–236

41. Greenberg RP *et al* (1981) Noninvasive localization of brain stem lesions in the cat with multimodality evoked potentials: correlation with human head-injury data. J Neurosurg 54 (6): 740–750

42. Herrmann D, Kunze K, Ágnoli A (1982) Orbicularis Oculi-Reflexe in der Differentialdiagnose der Hirnstammsyndrome. In: Struppler A (Hrsg) Elektrophysiologische Diagnostik in der Neurologie. G Thieme, Stuttgart New York, pp 246–247

43. Jellinger K (1968) Zur Neuropathologie des Komas und postkomatöser Encephalopathien. Wien Klin Wschr 80: 505–517

44. Jellinger K, Gerstenbrand E, Pateisky K (1963) Die protrahierte Form der posttraumatischen Encephalopathie. Nervenarzt 34: 145–159

45. Jellinger K, Seitelberger F (1970) Protracted post-traumatic encephalopathy. J Neurol Sci 10: 51–94

46. Jennett B, Teasdale G *et al* (1977) Severe head injuries in three countries. J Neurol Neurosurg Psychiat 40: 291–298

47. Jennett B, Teasdale G (1977) Aspects of coma after severe head injury. Lancet 23: 878–881

48. Jennett B *et al* (1979) Prognosis of patients with severe head injury. Neurosurg 4: 283–289

49. Jouvet M, Dechaume J (1967) Sémiologie des troubles de la conscience. Essay de classification. Rev Lyon Med 9: 961–968

50. Jouvet M (1969) Coma and other disorders of consciousness. In: Vinken PJ, Bruyn GW (eds) Handbook of clinical neurology, vol 3. North Holland, Amsterdam, pp 62–79

51. Kimura J (1970) Alterations of the orbicularis oculi reflex by pontine lesions. Arch Neurol (Chic) 22: 156–161

52. Kimura J (1971) Electrodiagnostic study of brain stem. Stroke 2: 576–586

53. Kimura J (1973) The blink reflex as a test for brain stem and higher central nervous system function. In: Desmedt JE (ed) New developments in electromyography and clinical neurophysiology, vol 3. Karger, Basel, pp 682–691

54. Kistler JP, Hochberg FH *et al* (1975) Computerized axial tomography: Clinico-pathologic correlation. Neurology 25: 201–209

55. Klug N (1982) Brain stem auditory evoked potentials in syndromes of decerebration, bulbar syndromes and in central death. J Neurol 227: 219–228

56. Klug N, Csécsei G, Rap Z (1982) Evoked potentials and blink reflex in clinical and experimental traumatic brain stem lesions. Excerpta Medica

57. Klug N (1983) Funktionsuntersuchungen des Hirnstammes im akuten Mittelhirnsyndrom unter Berücksichtigung vegetativer Größen während der Dezerebration. Habil-Schrift, Giessen

58. Klug N, Csécsei G (1985) Evoked potentials and brain stem reflexes. Neurosurg Rev 8 (1): 83–84

59. Klug N, Csécsei G (1985) Brain stem acoustic evoked potentials in the acute midbrain syndrome and in central death. In: Morocuti C, Rizzo PA (eds) Evoked potentials—neurophysiological and clinical aspects. Elsevier, Amsterdam New York Oxford, pp 203–210

60. Klug N (1986) Neurophysiological results (brain stem reflexes and evoked potentials) in primary and secondary brain stem disorders. In: Samii M (ed) Surgery in and around the brain stem and 3rd ventricle. Springer, Berlin Heidelberg New York London Paris Tokyo, pp 147–152

61. Laun A, Agnoli AL (1980) Brain stem haemorrhages. In: Pia HW, Langmaid L, Zierski J (eds) Spontaneous intracerebral haematomas—advances in diagnosis and treatment. Springer, Berlin Heidelberg New York, pp 196–201

62. Laun A, Agnoli AL, Schönmayr R, Villagrasa J (1983) Morphological and CT-findings in traumatic brain stem haemorrhages—a contribution of pathophysiology of primary and secondary lesions. In: Villani R (ed) Advances in neurotraumatology. JCS 612, Excerpta Medica, Amsterdam Oxford Princeton

63. Lyon LW, Kimura J, McCormick WF (1972) Orbicularis oculi reflex in coma: clinical, electrophysiological and pathological correlations. J Neurol Neurosurg Psychiat 35: 582–588

64. McNealy DE, Plum F (1962) Brain stem dysfunction with supratentorial mass lesions. Arch Neurol (Chic) 7: 10–32

65. Ongerboer De Visser BW, Goor C (1976) Jaw reflexes and masseter myograms in mesencephalic and pontine lesions. J Neurol Neurosurg Psychiat 39: 90–92

66. Ongerboer De Visser BW, Kuypers HG (1978) Late blink reflex changes in lateral medullary lesions. An electrophysiological and neuro-anatomical study of Wallenberg's syndrome. Brain 101 (2): 285–294

67. Ottaviani F *et al* (1986) Auditory brain stem and middle latency auditory responses in the prognosis of severely head-injured patients. Electroencephalogr Clin Neurophysiol 65 (3): 196–202

68. Overgaard J, Petersen K, Christensen S *et al* (1973) Prognosis after head injury based on early clinical examination. Lancet 11: 631

69. Pagni CA (1973) The prognosis of head injured patients in a state of coma with decerebrate posture. J Neurosurg Sci 17: 289–295

70. Pazzaglia P, Frank G, Frank F *et al* (1975) Clinical course and prognosis of acute post-traumatic coma. J Neurol Neurosurg Psychiat 38: 149–154

71. Perez-Dominguez E (1974) Les réflexes du tronc cérébral. Leur valeur dans l'étude des comas. Thèse, Université du Montpellier

72. Pia HW (1956) Die Einwirkung der Hirndrucksteigerung auf den Hirnstamm, ihre Klinik und Behandlung. Münch med Wschr 98: 1609–1612

73. Pia HW (1957) Die Schädigungen des Hirnstammes bei den raumfordernden Prozessen des Gehirns. Acta Neurochir (Wien) [Suppl] 4, Springer, Wien

74. Rappaport M *et al* (1981) Evoked potentials and head injury. 2. Clinical applications. Clin Electroencephalogr 12 (4): 167–176

75. Reeves AG, Posner JB (1969) The ciliospinal response in man. Neurology (Minneap) 19: 1145–1152

76. Rosenblum WI, Greenberg RP *et al* (1981) Midbrain lesions: frequent and significant prognostic feature in closed head injury. Neurosurg 9: 613–620

77. Rossi GF (1965) Some aspects of the functional organisation of the brain stem: neurophysiological and neurosurgical observations. Copenhagen, IIIrd Int Congr Neurol Surg. Excerpta Medica, pp 117–122

78. Sancesario G *et al* (1984) Prognostic evaluation of brain stem haematomas: the role of CT scan and brain stem auditory evoked potentials. Acta Neurol Scand 70 (6): 396–406

79. Schoenhuber R *et al* (1983) Neurophysiological assessment in progressive supranuclear palsy (letter). Ital J Neurol Sci 4 (3): 363

80. Schönmayr R (1973) Übersicht über die klinisch wichtigen Fremdreflexe des Menschen. Diss TU München

81. Schönmayr R, Laun A, Agnoli AL, Busse O (19821) Spontaneous brain stem lesions. CT-findings and clinical data in respect to morbidity. Advances in Neurosurgery, vol 10. Springer, Berlin Heidelberg New York, pp 47–49

82. Snyder BD *et al* (1981) Neurologic prognosis after cardiopulmonary arrest: IV. Brain stem reflexes. Neurology (NY) 31 (9): 1092–1097

83. Stewart WA, Litten SP, Sheehe PR (1973) A prognostic model for brain stem injury. Surg Neurol 1: 303–310

84. Struppler A, Dobbelstein H (1963) Elektromyographische Untersuchungen des Glabellareflexes bei verschiedenen neurologischen Störungen. Nervenarzt 34: 347

85. Tanaka Y, Kaga K (1980) Application of brain stem response in brain-injured children. Brain Dev 2 (1): 45–56

86. Tandon PN, Bhatia R, Banerji AK (1973) Vestibulo-ocular reflex and brain stem lesions. A clinico-pathological study. Neurology (India) 193–199

87. Teasdale G, Jennett B (1974) Assessment of coma and impaired consciousness. A practical scale. Lancet 2: 81–83

88. Turazzi S, Alexandre A, Bricolo (1975) Incidence and significance of clinical signs of brain stem traumatic lesions. J Neurosurg Sciences 19: 215–222

89. Uziel A *et al* (1982) Clinical applications of brain stem auditory evoked potentials in comatose patients. Adv Neurol 32: 195–202

90. Vapalahti M, Troupp H (1971) Prognosis for patients with severe brain injuries. Br Med J: 404–406

91. Verier A *et al* (1984) Evaluation clinique et pronostique d'un coma post-traumatique selon le nineau de souffrance du tronc cerebral. Sem Hop Paris, 60 (14): 1014–1019

92. Walshe FMR (1957) States of consciousness in neurology. Acta Med Belg 141–145

93. Werf AJM van der (1979) Brain stem injuries. European association of neurosurgical societies. Neurosurgical Course

94. Zee DS (1986) Brain stem and cerebellar deficits in eye movement control. Trans Ophtalmol Soc UK 105 (Pt 5): 599–605

95. Zuccarello M *et al* (1983) Importance of auditory brain stem responses in the CT diagnosis of traumatic brain stem lesions. AJNR 4 (3): 481–483

96. Zuccarello M *et al* (1983) Traumatic primary brain stem haemorrhage. A clinical and experimental study. Acta Neurochir (Wien) 67 (1–2): 103–113

Author's address: PD Dr. Robert Schönmayr, Department of Neurosurgery, University of Giessen, Klinikstrasse 29, D-6300 Giessen, Federal Republic of Germany.

Acta Neurochirurgica, Suppl. 40, 29–56 (1987)

Acute Direct and Indirect Lesions of the Brain Stem—CT Findings and Their Clinical Evaluation

Albrecht Laun

Department of Neurosurgery, University of Giessen, Federal Republic of Germany

Contents

Summary

Since the introduction of computer tomography (CT) (Ambrose 1973, Hounsfield 1973) it has become an essential instrument in the diagnosis of acute intracranial lesions. The precise analysis of the CT and in particular the evaluation of the basal cisterns[108], yields results which are already wellknown[56, 117], basically, from post-mortem investigations[39, 55, 81, 137] and clinical findings[98]. However, while these were retrospective analyses and results, serial CT examinations which are free of risk for the patient and can be used in correlation with the clinical neurological findings, allow important assertions *intra vitam*, as well as a definitely better assessment of the prognosis.

In addition to acute supratentorial lesions, acute and subacute infratentorial lesions are analysed and their clinical significance described. In this way, for the first time, the dynamics of the mechanical factors in raised intracranial pressure can be analysed. Important conclusions are drawn for the clinical management of the patients, and even some new indications for operation.

Analysis of the acute hyperdense brain stem lesions—pathognomonic for haemorrhages—allows for the first time the diagnosis and continous observation of traumatic and secondary haemorrhages caused by raised pressure, as well as spontaneous ones. As regards the mortality and morbidity, the results in this large series of traumatic and secondary haemorrhages are in striking contrast to previous analyses based on post mortem findings[49, 78, 79, 95, 96].

Acute hypodense brain stem lesions are not amenable to any definite pathogenetic classification—softening, inflammatory lesions, tumours and oedema must all be considered. With acute lesions we are only dealing with infarcts, which are only incompletely assessable in the computer tomogram, and their diagnosis must still depend on the clinical findings[31].

Secondary ischaemic lesions in acute raised intracranial pressure are identifiable in over 18% as infarcts which involve the entire territory of an artery. These additional space-occupying lesions are only survived by 11% of the patients. Hence the correlation which has been established between the basal cisterns and intracranial pressure is of great clinical significance.

From our own research group several reports on different aspects of raised intracranial pressure and lesions of the brain stem have appeared since 1979[3, 4, 5, 65, 66, 67, 68, 69, 108]. Similar analyses of partial aspects of the basal cisterns, have been published only recently[30, 51, 84, 87, 127] and came essentially to the same conclusions.

Keywords: Basilar artery occlusion; brain stem injuries; brain stem haemorrhages; brain stem infarcts; brain stem lesions; cerebral hypoxia; hypoxia in raised ICP; intracranial mass shift; locked-in syndrome; secondary infarcts in raised ICP; spontaneous haematoma of the brain stem.

Material and Methods

In 253 patients (pts) (190 male, 63 female) (Table 1) 877 computer tomographic examinations (766 without and 101 with contrast enhancement) were correlated with the progress of the neurological findings. All age groups were represented and the average age was 37.0 ± 19.5 (SD) years. In cranio-cerebral injuries (Fig. 1) the peak incidence was in the third, in lesions from other causes in the fourth decade. The division of the age-groups in relation to the Glasgow Coma Scale (GCS) (Table 2) showed that about 10% of the patients had a lower brain stem syndrome and about one half of the patients were in a stage of latent or manifest midbrain decerebration (GCS 04– 06) at the time of admission. The relation between the findings on admission (GCS) and the results of treatment (Glasgow Outcome Scale—GOS) show (Table 3) that 118 patients died, nine reached an apallic syndrome (persistent vegetative state—PVS), 25 survived with a severe, 42 with a slight neurological deficit and 59 without any significant deficit.

Table 1. *Case material*

Diagnosis	N		
Head injury	174	143	closed
		31	compound
Spontaneous haematoma	54	38	supratentorial
		16	infratentorial
Infarction	13	3	Carotid artery
		10	Basilar artery
Subarachnoid haemorrhage	11		
Haemorrhage into a tumour	1		

Table 2. *Age distribution in relation to the findings on admission*

Age GCS	0 – 14	15 – 19	20 – 39	40 – 59	>60	Σ
3	1	7	6	9	1	24
4 – 6	11	17	41	38	17	124
≥ 7	14	12	19	38	22	105
Σ	26	36	66	85	40	253

Table 3. *Total material.* Relation GCS/GOS

GOS GCS	Dead 1	PVS 2	Severely handicapped 3	slightly handicapped 4	good recovery 5	Σ
3	20	2	2	0	0	24
4 – 6	67	6	20	18	13	124
≥ 7	31	1	3	24	46	105
Σ	118	9	25	42	59	253

Methods

The ability to assess the cisterns was tested on 135 healthy volunteers. At the same time measurements of density were undertaken in these subjects with computer tomographs of the second and third generations. These showed variations in the spread which are at least twice as great as the values given by Hounsfield in 1973 and 1980.

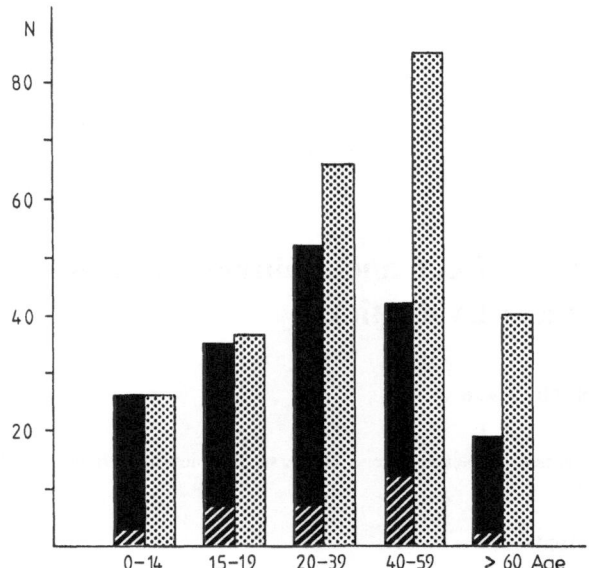

Fig. 1. Age distribution of the patients. Right stippled column: Total cases. Left column: Cranioventral injuries. Open head injuries—hatched

Because of their shape a computerized assessment of the basal cisterns is not reasonable. For this reason they are classified by purely visual subjective assessments into the categories unremarkable, narrow, obliterated and enlarged.

The assessment of the findings in the posterior fossa, more particularly in the region of the foramen magnum was restricted on account of the wellknown artifacts, associated specifically with the computer, with movement and in the border zones as well as the reduced resolution capacity. Brain stem infarcts, as for example in a Wallenberg syndrome, can not be shown by the CT. These particular problems will be discussed in detail later.

In each case the CT and the neurological findings were recorded on a questionnaire specially devised for the computer, with numerous individual features for each investigation. This also made possible a temporal correlation between one observation and another. In 56 patients continous recordings of the intracranial pressure (ICP, subdural), were available, which could also be correlated with the neurological and CT findings.

Results

In the literature about computer tomography the signs of herniation and of mass shifts in acute lesions, as well as the deformations and shifts of the cisterns in chronic space-occupying lesions[108], have been described by various authors, purely from the radiological viewpoint[21, 80, 86, 120, 133]. Osborn 1977/78[90, 91] described the special signs of descending and ascending herniation in tumours. As regards acute lesions there have been up to now articles by George 1981[36], van Dongen 1983[30], Toutant 1984[127] and Johnson 1986[51], which have examined particular aspects of the whole problem.

1. Supratentorial Lesions

1.1. Acute Supratentorial Lesions

The evidence of herniation into the tentorial hiatus is the increasing obliteration of the basal cisterns in the CT. In contrast to chronic lesions in which, depending on the size of the tumour, there is in 60%–80% of cases obliteration of individual cisterns[108], acute lesions always involve several cisterns. In acute lesions isolated hernias which can be seen in the CT scan develop in only 37% of cases[17].

Figure 2 shows a complete obliteration of the basal cisterns in an acute subdural haematoma after a cranio-cerebral injury. The brain stem can no longer be differentiated from the surrounding tissues and the subarachnoid space is no longer visible.

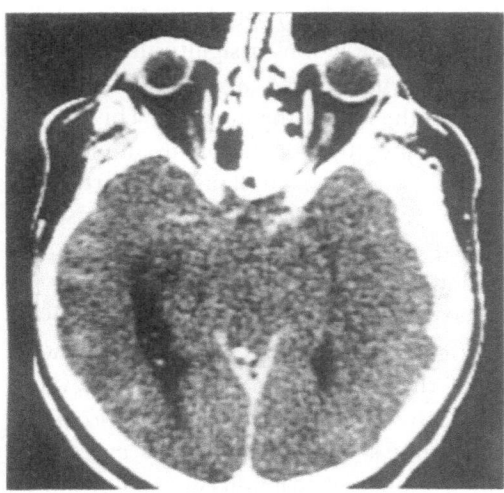

Fig. 2. Complete obliteration of the basal cisterns. Normal anatomical structures are no longer identifiable

Figure 3 is from an acute subdural haematoma with right temporal contusion and shows a right-sided uncal herniation with bilateral obliteration of the cisterna ambiens, right-sided obliteration of the cisterna pontis and narrowing of the cisterna quadrigemina on the right side. The brain stem is compressed from the right.

The *prognostic significance* of the analysis of the basal cisterns in the CT scans, in relation to the neurological findings is evident in Table 4. The analysis of the cisternal changes in the course of the temporal progress of the cisternal herniation covering time intervals of three days in each case (Table 5), provides

Table 4. *Correlation between findings on admission, course and condition of cisterns*

	GCS		GOS	
	03 – 06	≥ 07	1	≥ 2
All cisterns open	3	7	0	10
All cisterns oblit.	30	9	27	12
			p<	0,01

Table 5. *Correlation between cistern obliteration and mortality, at particular time intervals.* + <0.01 —not significant

Cisterns		Time interval			
		-2-0	1-3	4-6	7-9
interhemispheric cistern	R	-	-	-	-
	L	-	+	-	-
Cistern of Sylvian fiss.	R	+	-	-	-
	L	-	+	-	-
basal cistern	R	+	+	-	+
	L	+	+	-	+
interpeduncular cistern		+	-	-	+
Cistern of vein of Galen		+	+	-	+
quadrigeminal cistern	R	+	+	-	+
	L	+	+	-	+
ambient cistern	R	+	+	-	-
	L	+	+	-	-
pontine cistern		+	-	-	-

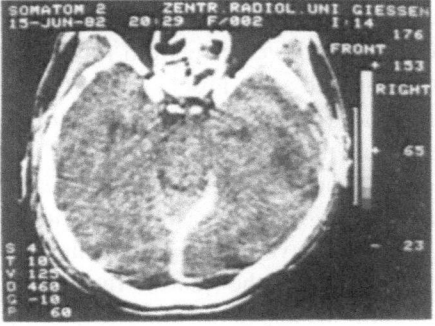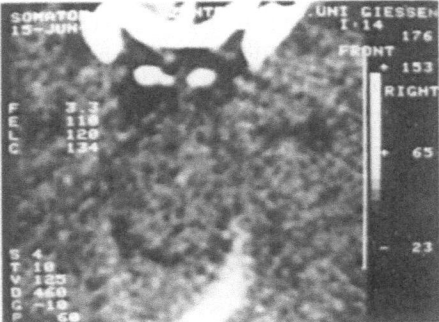

Fig. 3. **Right temporal herniation with narrowing of the basal cisterns**

further clues. Obliterations of the basal cisterns, Galeni, quadrigemina and ambiens show a significant connection with the mortality (p < 0.01). Up to the third day and, with the exception of the cisterna ambiens, also on the 7th–9th days. The relationship is not apparent in the case of the interhemispheric and the sylvian cisterns i.e. the cisterns are obliterated relatively early, nevertheless their significance for the prognosis is less. The constant absence of significance for the 4th–6th day corresponds to a marked cisternal obliteration even in surviving pts. Corresponding to that there is no prognostic significance in that period.

As a sign of mass shifts in supratentorial lesions which are predominantly unilateral, there is a difference in the changes in the cisterns on the two sides. An example of the findings in the basal cisterns is shown in Figure 4. Besides the group including all the cases (diagram a + b), the findings in the survivors (diagram c + d) and the fatal cases (diagram e + f) are shown separately. (The figures in the uppermost line of each separate diagram give the numbers of the respective CT recordings). As an example the basal cisterns demonstrate most markedly the difference between the survivors and those who died—it is less marked, but is nevertheless clearly shown by the cisterna ambiens. In the survivors the curves for the open and obliterated cisterns run almost parallel, but in the case of those who died are far apart. A significant result of the temporal course and the lateralised differences of the cisternal hernias in acute cases is the frequency of symmetrical obliterations and their prognostic significance. This is shown particularly in the cisterna basalis and cisterna Galeni, in which the fatalities in the former instance show about 40%–50% higher incidence of obliteration than the survivors, and in the latter it is three times more frequent—60% as against 20%.

With a complication-free course, cisternal hernias regress progressively from the eleventh day and after three weeks are no longer apparent.

In spite of the definite significance of herniation of the basal cisterns for the prognosis, deaths can occur as a result of failure of central regulation after an acute rise of intracranial pressure without herniation, or with slight narrowing of the cisterns (last time-interval, Figure 4). Of ten pts, four of whom were operated on, seven died from a clinically confirmed brain stem injury and three from peripheral causes.

Measurement of the density in healthy volunteers showed a variation in the brain stem density from 0.30 to 1.38% on the Hounsfield-scale. Measurements were also made in three groups of patients (Table 6): healthy

volunteers, those who were fit after a slight closed head injury and patients with an epidural haematoma with herniation. When these findings were grouped according to decades the results were only slightly significant. With CT examinations in proven herniations, in the fourth decade the density of the brain stem was higher and with a greater significance (p < 0.01) than it was in the 6th and 7th decades (p < 0.05).

In 60% of patients with lateral shifts, with compression of the ipsilateral ventricle there was a *hydrocephalic dilatation* on the opposite ventricle, more marked in the temporal and occipital horns. In comparison with patients without asymmetrical hydrocephalus, dilatation of the ventricular system provided no significant distinguishing feature, as regards morbidity and mortality.

An analysis of the *effects of age* showed in predominantly unilateral acute lesions a significant difference only for patients up to the age of 20 years, in comparison with the rest of the material. This applied particularly in obliteration of the basal, ambient and interpeduncular cisterns, as well as for partial obliteration of the galenic and interpeduncular cisterns. With lateralised mass shifts with ipsilateral narrowed and contralateral obliterated cisterns those up to 20 years of age have likewise a more favourable prognosis.

The investigations into any correlation between obliteration of the cisterns and the patients level of consciousness, as well as of individual *neurological syptoms* were only significant in a few cases. The absence of the eye-opening reactions was significant for all cisterns, if they were obliterated (p < 0.01), but there was no correlation in the absence of verbal contact. In the brain stem syndrome with motor decerebration or with absent motor reactions there was a significant correlation with obliteration of the basal and galenic cisterns and the ambient cistern bilaterally. The corresponding clinical investigations for all cranial nerve- and brain stem-reflexes, as well as spontaneous motor function, tone and automatism were only significant in the case of absent direct and indirect pupillary light reactions. There was a positive correlation for dilated and irregular pupils as well as an absent corneal reflex with obliteration of the cisterna Galeni, cisterna pontis, the cisterna interpeduncularis and to some extent also for the cisterna quadrigemina.

The *intracranial pressure* was measured subdurally in 56 pts. In correlating the average pressure value over a period of 24 hours with the cistern findings in complete herniation on the one hand, and with normal or narrow cisterns on the other, and with the adoption

supratentorial lesions
basal cistern

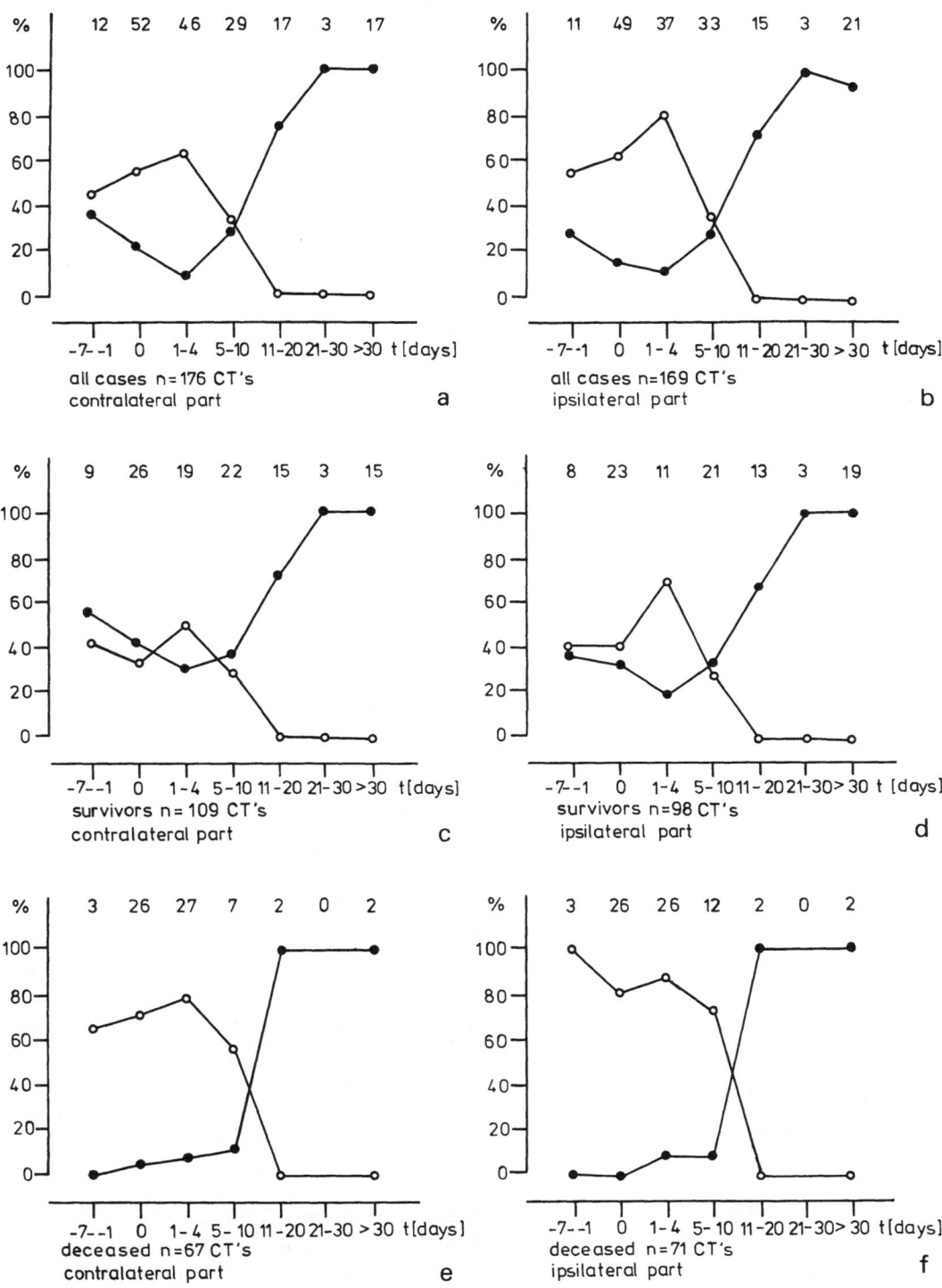

Fig. 4 a–f. Extent and regression of cistern obliteration. Circles: obliterated cisterns. Dots: open cisterns

Table 6. *Brain stem density in patients with and without herniation, in relation to decades (× p < 0,01, × × p < 0,05)*

Age (yrs)		Healthy volunteers	Minor closed head injury without neurol.deficit	epidural haematomas with herniation	
0 - 9	average	26,1	26,3	-	
	SD	3,6	5,8	-	
	N	11	10		
10-19	average	27,1	28,3	25,3	
	SD	5,2	5,1	11,0	
	N	10	9	9	
20-29	average	27,6	28,0	28,0	
	SD	5,7	5,6	5,8	
	N	10	10	8	
30-39	average	27,7	27,9	37,0	x
	SD	4,9	4,3	8,7	
	N	10	9	8	
40-49	average	22,8	29,9	28,5	
	SD	3,1	5,6	9,6	
	N	10	10	9	
50-59	average	26,6	24.8	30,4	xx
	SD	2,5	3,3	6,0	
	N	10	10	8	
60-60	average	26,0	25,3	32,8	xx
	SD	3,8	6,2	6,5	
	N	10	9	8	
70-70	average	26,0	24,1	-	
	SD	3,4	3,5	-	
	N	9	10		

of a critical level for the intracranial pressure, either at 15 or 30 mm Hg, significant relationship at the 99% or the 95% level were found (Table 7). Only in one solitary

Table 7. *Correlation between herniation and intracranial pressure. Critical limit at 15 mm Hg or at 30 mm Hg*

		15 mmHg		30 mmHg	
		Chi^2	significance	Chi^2	significance
Interhemispheric cistern	R	10,85	p< 0,01	7,88	p< 0,05
	L	5,63	-	6,51	p< 0,05
basal cistern		2,28	-	2,7	-
cistern of Sylvian fiss.	R	11,5	p< 0,01	4,13	p< 0,05
	L	8,37	-	5,0	-
cistern of vein of Galen		4,81	-	6,39	p< 0,05
quadrigeminal cistern	R	3,34	-	7,06	p< 0,05
	L	1,77	-	9,66	p< 0,01
ambient cistern	R	3,14	-	5,43	-
	L	1,29	-	9,65	p< 0,01
interpeduncular cistern		9,09	p< 0,05	12,4	p< 0,01

patient with a raised intracranial pressure no correlation was found in CT. In this way computer tomographic conditions are shown, in which an intracranial pressure of over 30 mm Hg is possible, namely: complete obliteration of the basal cisterns by herniation, their tamponade with blood, or a haemorrhage in the aquaeduct or the fourth ventricle, as well as obliteration of the infratentorial subarachnoid space.

1.2. Decerebration Without Herniation

Eight head-injured patients (Table 8) between 11 and 29 years of age were decerebrate or had evidence of an incomplete brain stem syndrome without any marked herniation being detectable in the CT. In four out of five cases the measurement of the brain stem density showed a variable but marked reduction within the first ten days, with a subsequent return to normal within

Table 8. *Decerebration with minimal lesions demonstrated in CT*

Decerebration without detectable lesion and normal
cisternal appearance in CT

	Age/Sex		GOS	ICP
1	17	m	3	-
2	21	m	3	not elevated

Decerebration with minimal CT findings and normal cisterns

	Age/Sex		GOS	Lesion	Cisterns
3	13	f	5	Blood in the subarachnoid space above the left tentorial ridge, IVH	quadr.cistern and left ambient cistern narrowed
4	29	f	3	left front.EDH bitemporal small contusions	quadr.and ambient cisterns narrowed
5	11	f	3	small bitemporal contusions IVH	quadr. and ambient cisterns narrowed
6	19	m	3	bifrontal and left temporal contusional foci	ambient cistern narrowed
7	19	m	3	small left SDH (max.diam. 6 mm) subarachnoid haemorrh. over the left tentorium	
8	19	m	4	rt. frontal contusion	

three weeks. Clinically all patients were primarily unconscious, two patients were again responding after two days but the other six remained comatose for 2–4 weeks. All patients survived, a thirteen-year-old girl with no neurological deficit, a 19-year-old with slight deficit (GOS 4) and the remainder with severe neurological deficits (GOS 3).

In comparison with all other decerebrate pts, those with no signs of herniation have a statistically better chance of survival (p < 0.01).

1.3. Discussion of Supratentorial Lesions

The examination of the cisterns with determination of the degree of obliteration must be emphasized as an important factor in the assessment of the prognosis of *acute* lesions. This had confirmed our earlier observations[3, 4, 5, 66, 67]. With this analysis of the course of obliteration of the cerebral cisterns it has become possible for the first time to gain an *intra vitam insight* into the dynamics of herniation and the shifts of the brain stem, on a morphological basis. George[36] assessed the cisterns in relation to the size of the ventricles, in order to be able to distinguish primary from secondary brain stem lesions in the CT. Narajan *et al.*[88] were unable to make any useful prognostic statements, on the basis of CT findings alone; they have, nevertheless, not included the cisterns in their considerations. On the

other hand van Dongen[30] made precise statements by judging the type of lesion in relation to the extent of cisternal obliteration. They likewise distinguish three degrees of deformity: free, partly distorted (narrow) and obliterated. This global estimation is reasonable, but it has the disadvantage, that the distinctive significance of the cisterns is neglected. Toutant[127] confirmed this association between the cisterns as shown in the first CT and the outcome. This association was shown to be strong after adjusting for the Glasgow Coma Scale.

The itemised assessment at different times makes clear that no pronouncement regarding the prognosis is possible between the fourth and sixth days, which is thus contrary to the assertions of van Dongen. The extent of the obliteration increases in all the cisterns up to the fourth to sixth days (Figure 4) and after that gradually decreases. In the case of a review between the seventh and ninth days, all the basal cisterns show a significant correlation with the mortality. This gives the impression, that with an obliteration of the cisterns existing over this period, apparently irreversible structural damage to the brain stem develops. From the 10th to the 20th day onwards the cisterns are more frequently open than obliterated; after three weeks there are no longer any signs of herniation. Exceptions are patients with secondary complications, as for example a meningo-encephalitis after a compound head injury.

The temporal course first pointed out in the CT corresponds to the individual observations of central disturbance of autonomic regulation[12, 70, 75, 113] which likewise improved after 2–3 weeks. These find, at least partly, their morphological correlation in the pattern of obliteration of the cisterns.

The consequences of the herniation have been discussed in detail from the point of view of the morbid anatomy e.g. the compression of the vein of Galen[118], as well as the diversion of the venous drainage from the superficial pontine and intercrural veins, so that it is no longer into the vein of Galen, but into the veins of the sylvian fissure[63]. This readjustment of the circulation could be an explanation of the significant increase in the density of the brain stem in herniation, which is identifiable in the fourth, sixth and seventh decades. In the third and fifth decades the average density in herniation is also higher, nevertheless not statistically significant. The considerable extent of the variations in density in the controls may be a further explanation for this.

Individual neurological symptoms show a significant correlation with obliterated cisterns. In the Glasgow Coma Scale[124] the level of consciousness is classified according to three factors, verbal contact, eye opening and motor responses; significant correlations are found with the last two of these. Other significant relationship exist besides the absent light reaction, such as the size of the pupils and the corneal reflex which was also described by van Dongen.

The not uncommon involvement of the cisterna pontis in the case of acute supratentorial lesions with positive correlations is the expression of cranio-caudal transtentorial herniation and represents an additional danger factor. This is seen particularly in the temporal course at the onset and with constant displacement between the seventh and ninth day. Any preference for the obliteration of individual cisterns in relation to the site of the hemisphere lesion could not be identified. This may be because of the frequent occurrence of multifocal lesions. Two-thirds of the whole material had multifocal and diffuse cranio-cerebral injuries and one quarter had bilateral lesions. A further reason could be the widely spaced CT examinations in the presence of rapidly progressing acute lesions.

With the investigations into the correlation with ICP, for the first time significant relationship were discovered between the CT and the intracranial pressure. By adopting a critical level of 30 mm Hg and observing the obliteration of the infratentorial subarachnoid space, a positive correlation was found in 95% of cases. Other observers[9, 40, 123] by assessing the size of ventricles, lateral mass displacement and the nature of the hemisphere lesion found no unequivocal correlation. An increase in the ICP over 30 mm Hg showed three possible correlates in the CT:

a) Obliteration of the basal cisterns or their tamponade with blood.

b) Intraventricular blood in the aquaeduct and fourth ventricle.

c) Obliteration of the infratentorial subarachnoid space.

In recent times Lobato and his associates[74] confirmed this in their statement that ICP measurements are not necessary in head-injured patients with a normal CT.

Herniation without any impairment of consciousness is very rare in acute supratentorial lesions. Chronic types of progress are distinguished by a primary compensation and adaptation and, even with marked herniation need not themselves be associated with clouding of consciousness and other midbrain syndromes[108]. The only solitary case observed with an almost complete herniation without impairment of consciousness underlines the possibilities for compensation in younger subjects. It says nothing however about the pathogenetic significance of cisternal hernias as an expression of the mass displacement and raised intracranial pressure. Establishment of the diagnosis calls for the appropriate treatment.

Pts under 20 years of age have a significantly more favourable prognosis, even in the case of lesions with lateralised mass shifts. Recently Berger[14] was able to demonstrate a better outcome for children up to 10 years of age compared with adolescents (11–17 years). In patients with lateralised shifts 60% show in CT a hydrocephalic dilatation of some or all of the contralateral ventricles.

Stovring[120] found a dilatation of the contralateral temporal horn in over 16% of cranio-cerebral injuries. His assumption that this dilatation proves the presence of transtentorial herniation into the cisterns with an aquaeduct stenosis and that a normal temporal horn proves its absence, cannot be confirmed either morphologically or by our CT findings[108]. The different figures are explained by the fact that in our material we have considered exclusively only those injuries with verified lateralised mass shifts. The lateral shift has resolved in more than two-third of the cases within ten days. With this reduction of the shift the asymmetrical dilatation of the lateral ventricles also regresses.

In view of this analysis of the cisterns, any hydrocephalic dilatation of the ventricles is to be regarded as rather a late sign of the herniation. It only occurs, if as a result of a mass shift the contralateral foramen of Monro or the trigone is obliterated or compressed. As a result of this additional increase in volume the intra-cranial pressure rises. Correspondingly, the mortality rate rises, if only slightly (not significant).

Summary

All the basal cisterns show, when they are obliterated a highly significant correlation with the mortality. By detailed analysis of the findings in the cisterns and their correlation with the neurological findings as well as their progress, significant relationships have been worked out. The shape and reaction of the pupils, the corneal reflex and absence of the eye opening reaction as well as the motor response in coma, show a significant relationship with complete obliteration of the basal cisterns.

Pts under 20 years of age have a better chance of survival than older pts. The regression of the cisternal obliteration has partly occurred after ten days and after about three weeks is complete. After this time the mechanical factor no longer plays any part and any neurological defects must be regarded as due to structural damage.

2. Infratentorial Lesions

Infratentorial space-occupying lesions with an acute course of hours or a subacute course of a few days lead, as a result of local damage and compression, to extensive mass shifts with damage to the upper cervical cord, the pons and the midbrain. These shifts can be identified in the CT.

In a case of spontaneous haematoma of the cerebellar hemisphere (Fig. 5) there was a marked increase in the infratentorial pressure. Only the supratentorial portion of the cisterna pontis is free, while the other infratentorial CSF spaces are obliterated. The supratentorial sections of the midbrain are flattened and deformed in a sagittal plane. Corresponding to the ascending herniation the cisternae Galeni et quadrigeminae are obliterated. In this slice the cisterna ambiens is free above the tentorium (Fig. 5a). The coronal reconstruction (Fig. 5b) confirms the cranial herniation with filling in of the tentorial hiatus and shows in detail the obliteration of the right infratentorial portion of the cisterna ambiens and its narrowing above the tentorium. The cisterna magna is almost completely obliterated. The two tonsils, the right clearer than the left, are caudally displaced and visible below the foramen magnum (Fig. 5e).

The analysis of the CT pictures shows that direct operative relief substantially hastens the regression of the herniation (Fig. 6).

CT analysis can not only affect the indications and timing of an operation but also enables us to express an opinion about the prognosis. The acute lesions can be distinguished from the subacute on the basis of their clinical course i.e. the time from the onset of the first symptoms until the deterioration of consciousness.

2.1. Acute Infratentorial Lesions

Material

In this group of patients were included 15 craniocerebral injuries and two spontaneous cerebellar haematomas. Of the 15 patients with suboccipital head injury there were: five with cerebellar contusion without any CT evidence of supratentorial injury, five cerebellar contusions of whom three had small supratentorial subdural haematomas and two had small contusions, one with a diffuse lesion in the posterior fossa with supratentorial intraventricular bleeding and four posterior fossa epidural haematomas, one with no supratentorial injury, one with a small temporal and two with an intraventricular haemorrhage.

All ages were represented. Pts under 30 years have a significantly better chance of survival (Fischer test p = 0.06).

Clinical Findings

All seventeen patients showed clouding of consciousness and 13 were decerebrate. In four cases there was a primary respiratory arrest and three times a cardiac arrest. Eleven had pathological light reactions and in eight the corneal reflex was absent (Table 9). Nine of the 13 decerebrate patients died. Only one 13 years old child survived a decerebration syndrome without any significant deficits: in this case only the cisterna Galeni was obliterated and the cisterna pontis narrowed.

Table 9 *Frequency of clinical symptoms in acute infratentorial lesions*

Symptomatology	
Disturance of consciousness	17
Decerebration	13
Primary respiratory arrest	4
Primary cardiae arrest	3
Pathol.or absent light reactions	11
Absent corneal reflex	8
Pupillary abnormalities	4
Deminished corneal reflex	3
Abducens paresis (one bilateral)	3
Facial paresis	2
Increased saliva secretion	2
Lower cranial nerve disturbances	2
Hemiparesis	4
Cerebellar symptoms	3
Hypotonus	4

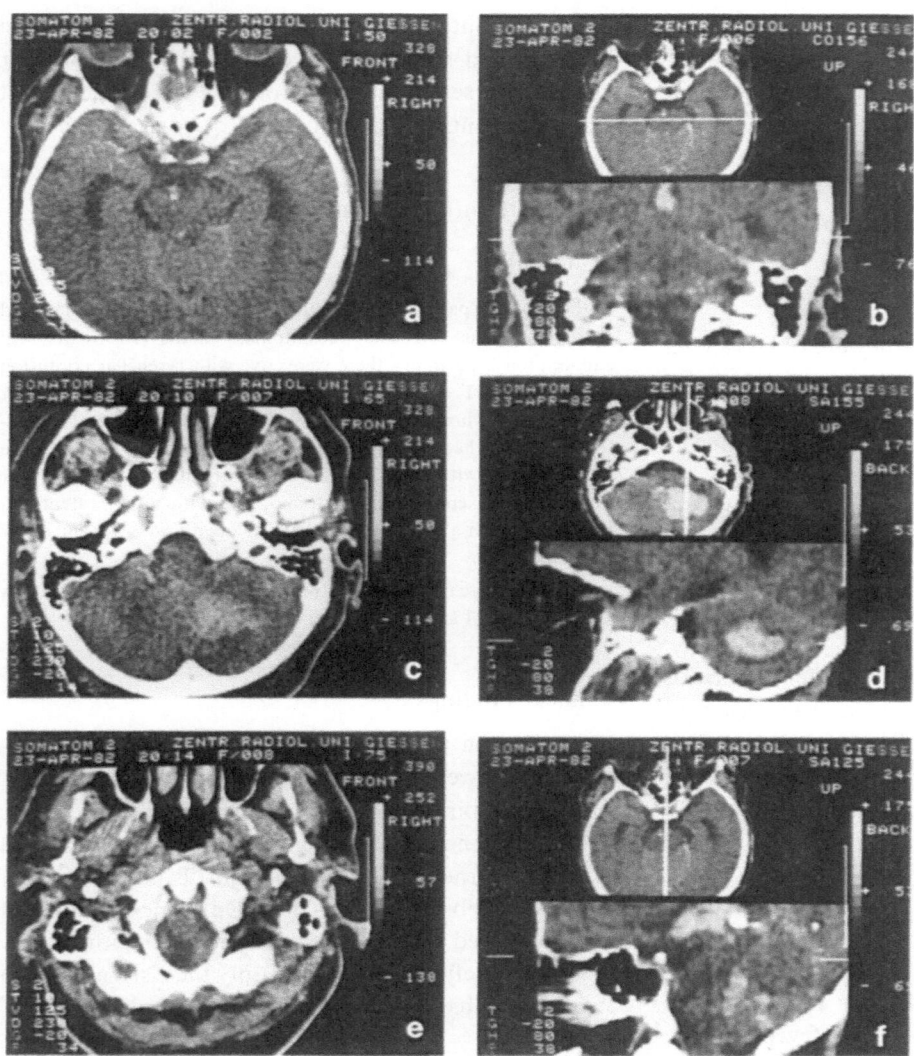

Fig. 5 a–f. Ascending herniation with cerebellar haematoma. For details—see text

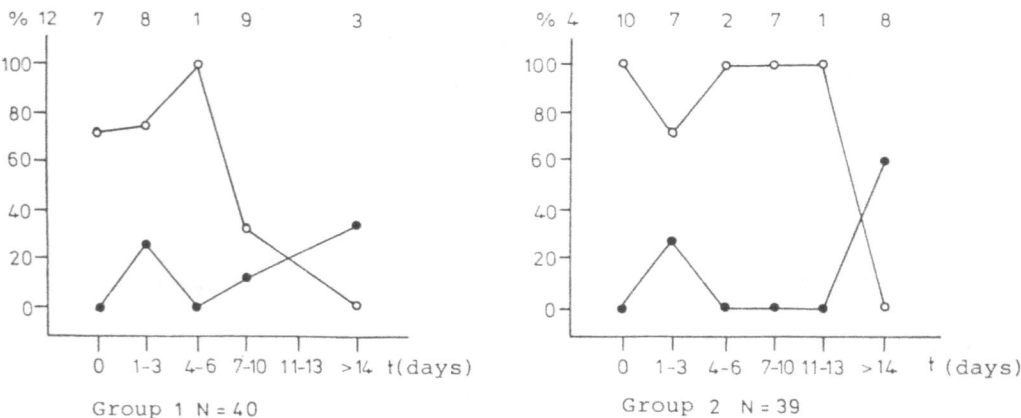

Fig. 6. Space-occupying lesions in the posterior fossa. Cistern of the vein of Galen. Obliteration and demonstration of the cisterna Galeni at different times. Group 1: Decompressive suboccipital craniotomy. Group 2: Without decompressive craniotomy

CT Findings

Tonsillar Herniation:

In five patients the tonsils were displaced downwards in the first CT. In all the other patients the foramen magnum could not be examined.

Upward Herniation:

In all patients there was a greater or lesser degree of upward herniations. The cisterns were completely obliterated in eleven pts, of whom only two survived. Of the remaining six patients with partially obliterated cisterns, five survived (Fischer test significant, p = 0.018). All eight patients who were decerebrate on admission and had obliterated cisterns died. Five patients with normal pupil, light reaction and corneal reflex survived. The exception was one patient who died as a result of alcoholic liver disease.

Compression of the Brain Stem:

Compression of the brain stem was apparent in eight pts, in seven of whom the cisterna pontis was obliterated. In the eight there was an epidural haematoma with partly open cisterns. Six of these patients died. Two had epidural haematomas and who were conscious on admission made a full recovery. Decerebrate patients with compression of the brain stem all died, while of those without compression 50% survived (8 : 4, Fischer test significant p = 0.098). Only two of those eight patients with brain stem compression showed regular and active pupils and corneal reflexes. Signs of brain stem compression in acute infratentorial lesions must be regarded as an additional unfavourable prognostic sign.

Hydrocephalus

In eight patients there was a hydrocephalic dilatation in the first CT scan. In seven cases the infratentorial cisterns were obliterated and in one case they were filled with blood with exception of the cisterna pontis. Only one of these patients survived with minor neurological deficits. He had suffered head injury with a suboccipital fracture and the CT showed a very small subdural haematoma over the left cerebellar hemisphere and an intracerebellar haemorrhage on the right. On admission he was not decerebrate, the pupils, light reflexes and corneal reflexes were normal. Continuous drainage proved ineffective and it was only after suboccipital craniotomy on the fifth day that there was a definite improvement with lightening of the conscious state.

When hydrocephalus develops in association with an acute space occupying lesion in the posterior fossa it must be regarded as an additional unfavourable factor.

Results

By comparing the patients with complete obliteration of the cisterns (ascending herniation) with those with only partially obliterated cisterns, the latter have a significantly better chance of survival. No patients in a decerebrate state with compression of the brain stem survived while four out of eight of those with no signs of brain stem compression did survive. Because of that not only complete ascending herniation with obliteration of the cisterns, but also brain stem compression, must be regarded as an additional unfavourable prognostic sign in acute infratentorial lesions. The same holds true for the development of dilatation of the ventricles in the case of obliteration of the infratentorial subarachnoid space, or its haemorrhagic tamponade.

2.2. Subacute Infratentorial Lesions

Included in this group were ten patients with cerebellar infarcts, four patients with spontaneous hypertensive haematomas and one with cranio-cerebral injury and cerebellar contusion and one with subarachnoid haemorrhage. The patients were between 30 and 79 years old.

Clinical Findings

The prodromal syndromes lasted two to seven days (once fourteen days), on average 3.8 days. Twelve out of 16 patients showed disturbances of consciousness and were decerebrate. The most frequent symptoms are listened in Table 10. The three patients with decerebrate symptoms survived with minimal neurological deficit. One with cerebellar infarct, one with spontaneous cerebellar haemorrhage and one with subarachnoid haemorrhage died.

CT Findings

Tonsillar Herniation

A caudal displacement of the tonsils in the foramen magnum was always apparent except in one case. Hence it was not possible to relate this CT finding to any particular neurological symptoms.

Upward Herniation

The infratentorial cisterns surrounding the brain stem were obliterated in ten patients and narrowed in six. All patients showed evidence of this upward herniation. The Galenic cistern was obliterated in 12 cases and narrowed in four. The same was true for the quadrigeminal and ambient cisterns. The analysis of the neurological symptoms did not allow any distinction to

Table 10. *Frequency of clinical symptoms in subacute infratentorial lesions*

Symptoms

Disturbances of consciousness	12
Decerebration	3
Respiratory disturbances	2
Impaired ocular movements	9
Path.light reaction	3
Pupillary changes	1
Absent corneal reflex	4
Abducens paresis	1
Facial paresis	5
Impairment of lower cran. nerves	5
Hemiparesis	4
Cerebellar symptoms	11
Headaches	12
Nausea	7
Vertigo	8
Vomiting	6

be made between those with obliterated and those with narrowed cisterns.

Compression of Brain Stem

Compression of brain stem was apparent in the CT in 13 pts. In ten the cisterna pontis was obliterated and in three significantly narrowed. In one of these there was local deformation of the brain stem. No correlation could be shown between the CT signs of compression and any particular neurological symptoms. Because of the subacute course of this brain stem compression some compensations occurred and fewer neurological symptoms developed.

Hydrocephalus

In 13 patients there was hydrocephalic dilatation of the ventricles in the first CT. No difference in neurological symptoms or between morbidity and mortality was apparent between those patients with and without hydrocephalus.

With the acute lesions there are particular neurological symptoms and an effect on the course of the illness, associated with the varying extent of the herniation. In the case of the subacute lesions this is no longer possible on account of the compensation mechanism which have already developed within one to seven days. That this is probably a question of a very labile compensation is shown by the very sudden decompensation associated with iatrogenic upset of this pathological balance, whether it be due to puncture of the ventricle (continuous drainage) or due to lumbar puncture. After such measures it is possible to show in the CT (Fig. 7) an increase in the herniation, which runs parallel with an increase of the neurological irritability or deficits, right up to the decerebrate state. There is a good chance to save such patients by immediate direct operative decompression, as shown in our own three cases who survived with minimal neurological deficits.

2.3. Discussion of Infratentorial Lesions

It was Osborn[91] who, on the basis of seven observations first described the signs of caudal and cranial herniations in the CT picture. The findings were verified in four cases at operation, in one by angiography and in another by the further progress of the case, where regression of the condition was shown in the CT. In comparison with the static description, a dynamic study with the use of sagittal and coronal reconstruction

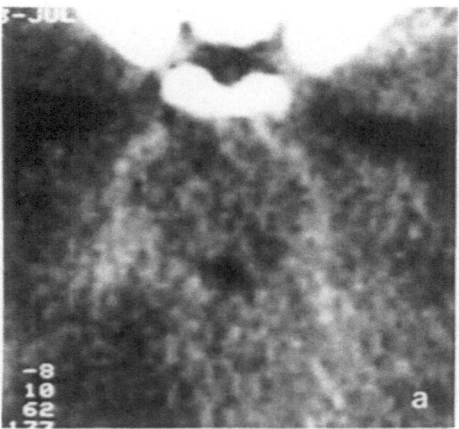 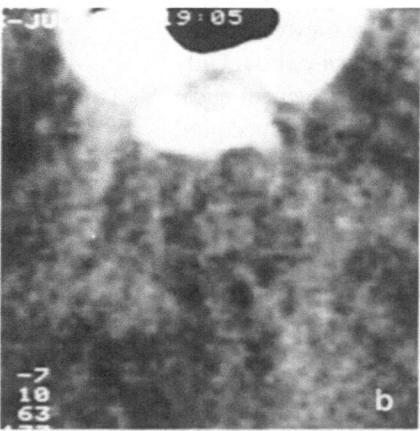

Fig. 7. a) Ascending herniation in a left cerebellar infarction, aquaeduct visible, cisterns obliterated. b) 6 hrs after ventricular drainage. Aquaeduct compressed, cisterns obliterated

yields significantly more information. The anatomical information obtained in this was correlated with the pathological changes in infratentorial space-occupying lesions and with the neurological findings which have been collected[58]. The confirmation of the caudo-cranial transtentorial herniation and the caudal tonsillar herniation through the foramen magnum by the sophisticated CT-technique is decisive. The dangers of lumbar puncture in tonsillar herniation are well-known and induced Spatz and Stroescu[117] to recommend ventricular puncture. In three cases with cerebellar infarctions, where ventricular drainage had been established, we were able to demonstrate progressive deformity and obstruction of the aqueduct, thus for the first time verifying in the CT the dangers of ventricular puncture with increase of the transtentorial midbrain compression. In these three cases a decision was made to carry out an urgent suboccipital craniotomy. This produced a striking reversal of the clinical deterioration in all cases. Recently Cioffi[20] confirmed that "it seems likely that ventricular drainage may account for an upward transtentorial herniation". His and our own findings are in agreement with those of Tsai[129]. However, we are unable to confirm his idea that patients can only survive this stage in a "persistent vegetative state" (PVS). In acute localised lesions, the indication for operation is quite clear and it must take place as quickly as possible[119]. The differentiation of subacute lesions and their computer-tomographic analysis shows that, even in these cases, operative intervention may be indicated.

With subacute lesions a hydrocephalus develops more frequently. On account of the compensation mechanism it has no decisive influence on the subsequent progress. On the other hand, in the case of acute lesions the development of hydrocephalus clearly worsens the prognosis. Hinshaw[42] maintained that direct intervention is not necessary in the case of cerebellar infarcts and recommend drainage of the associated hydrocephalus. Tomaszek[126] recommends angiography and Doppler sonography but in acute deterioration this makes no sense. Lui[76] favours operative treatment in comatose patients with spontaneous cerebellar haemorrhage. However, it is not justified, according to the foregoing results, to hope for a favourable course as there are neither clinical nor CT signs, which allow a definite opinion regarding the further progress[10, 16], although an improvement in cerebellar infarcts is possible and has been described[43, 110, 129].

Because of the limited power of resolution in the posterior fossa it is not possible, for example in cerebellar infarcts, to recognise any involvement of the brain stem. This may explain the variable postoperative courses in subacute infratentorial lesions.

Summary
The computer-tomographic picture of the acute and subacute space-occupying lesions in the posterior fossa has three characteristics:

1. Narrowing or obliteration of the cisterna Galeni and the infratentorial portion of the basal cisterns which surround the brain stem.

2. Symmetrical or asymmetrical compression of the brain stem in a sagittal direction, with obliteration of the cisterna pontis.

3. Prolapse of the tonsils into the foramen magnum.

In spite of the occurrence of compensation mechanisms in the case of a subacute course, direct operative relief is indicated in localised lesions, especially when neurological deterioration occurs, as it is the only measure which eliminates the cranial as well as the caudal herniation.

3. Direct Changes in the Brain Stem

3.1. The Hyperdense Lesion, Brain Stem Haemorrhage/Haematoma

Acute hyperdense lesions in the plain CT are, as in the other parts of the CNS, haemorrhages. The finding is pathognomonic and allows for the first time *intra vitam* the diagnosis, with evidence of the site, the size and the extent of the lesion, as well as control of the progress.

3.1.1. Traumatic Direct and Indirect Brain Stem Haemorrhage

Traumatic brain stem haemorrhages arise as a direct effect of rupture of vessels and/or through a secondary rise in intracranial pressure with cistern herniations and mass displacements, as in acute traumatic and spontaneous space-occupying lesions.

The morbid anatomy of primary traumatic brain stem injuries has been described[48, 72, 82, 121]. The possibility of surviving them is questioned on account of the lack of definite *intra vitam* evidence[78, 79, 130]. All these previous analysis were based solely on experiments[114] and post mortem studies[19, 24, 131].

In secondary haemorrhages, the result of herniation and pressure, many factors have been considered as responsible, including: —venous congestion and thrombosis[38], —compression of the vein of Galen[11, 118], —disturbances of arterial circulation in the small pontine vessels[35, 134] and also in the basilar artery. The discussion has remained controversial for a long time[18, 61, 99]. Zülch[137], apart from the significance of the rise in intracranial pressure, regards the venous and arterial congestion as the essential pathogenetic mechanism in the caudal displacement of the brain stem.

The *material* analysed from all age groups, includes 32 verified traumatic primary and secondary brain stem haemorrhages caused by herniation. Doubtful findings and secondary haemorrhages in chronic space-occupying lesions, such as tumours, were excluded. The frequency is about 4% of the cranio-cerebral injuries treated as in-patients.

Results

By observing the shape of the haematoma and of the cisterns in the CT the cases may be divided into three groups:

Group A: Unilocular, clearly demarcated haematomas, with open basal cisterns (Table 11).

Group B: Haematomas blurred in outline, also multilocular developing ones, with complete obliteration of the basal cisterns (Table 12).

Group C: Haematomas, unilocular or multilocular, with clear or blurred outlines, with incomplete obliteration of the basal cisterns (Table 13).

Haematomas in *group A* were identifiable in the first CT. These six haematomas were opposite the tentorial margin, at the level of the lateral sulcus of the midbrain or in the midline. They were clearly demarcated and had a maximum axial extent of 10 mm. Their density was over 50 HU in 3 of 6 cases. It was characteristic for these patients that in the CT there were minimal or no lesions detectable elsewhere (Table 11).

In *group B* (Table 12) are grouped patients in whom the brain stem haemorrhages developed as multilocular with completely obliterated cisterns. The haemorrhages were frequently vaguely outlined and were identifiable rather more caudally in the pontine region, or else extended as far as this. They were situated in the midline and often extended towards the paramedian area, at times involving the entire cross section. In three of 13 cases they were only identified in a second CT.

In *group C* (Table 13) are grouped brain stem haemorrhages which simultaneously show features of group A and B, in whom however the basal cisterns were only partially obliterated. From their location some haemorrhages fall into group A, others into group B, but both types of haemorrhage can also exist side by side. Also in two patients in this group the haematomas were only detected at a follow-up CT examination. The density was between 30 and 50 HU.

Table 14 shows the density of the brain stem haematomas in the various groups. With a variation extending from 0.39–0.69%, the absolute value of the density (Table 14) gives only a hint for the differentiation of brain stem haematomas. Haematomas with a value of over 50 HU are in group A, densities below 40 HU are more frequently in group B. In patients in group A the density diminishes during the subsequent course and in group B there was an increase in density on three occasions. None of the survivors showed an increase in density during the course of the illness.

A more significant finding, well recognisable in the CT, is the mass shift. By classifying groups A–C from the point of view of mass shifts (Table 15) it can be shown that among those with a mass shift only one patient survived and he ultimately showed an apallic syndrome. All the other patients died. Only patients without a mass shift can survive an established brain

Table 11. *Brain stem haemorrhages group A*. Abbreviations Table 11–13: MS Midline shift; IVH intraventricular haemorrhage; SDH subdural haematoma; EDH epidural haematoma; cont. contusion; lf left; rt right; bl bilateral

No	Age/Sex	GCS	GOS	Interval since the accident	CT findings	Remarks
1	5/m	15	5	10 d	none	hemophilia A
2	7/m	07	5	6 d	none	
3	8/m	04	3	2,5 hrs.	subocc.fract.concussion in the vermis, no supra-tent. lesion	Galenic cistern obliterated
4	24/m	04	3	2 hrs.	pinpoint haemorrhage in the left fr. region	
5	40/f	15	4	6 d.	none	anticoagulant.ther. because of valvuloplasty
6	56/m	05	1	21 hrs.	haematoma in the Sylvian fissure	extracerebral cause of death

Table 12. *Brain stem haemorrhages group B*

No	Age/Sex	GCS	GOS	Interval since the accident	CT findings	Remarks
1	11/f	04	1	4 hrs.	MS lf→rt diffuse swelling	occ.infarction bl brain stem hypodensity
2	22/m	03	1	1 hr.	acute SDH lf, MS lf→rt bitemp.mult.cont.,IVH	
3	31/m	03	1	2 hrs.	acute SDH lf. MS lf→rt cont.lf.fr., IVH	occ.infarction bl
4	39/m	04	1	3 d	subacute SDH lf.MS lf→rt IVH	occ.infarction lf
5	39/m	03	1	8 hrs.	bilat.EDH occ.and subocc. rt.fr.cont., IVH	
6	42/f	03	1	24 hrs.	subac.SDH rt , MS rt→lf	occ.infarction rt
7	43/m	03	1		temp.cont. IVH,basal ganglia haem. rt	
8	46/m	11	1	2 hrs	acute SDH rt, MS rt→lf postop. EDH lf ∅ MS	occ.infarction rt
9	53/m	03	1	14 hrs.	acute SDH rt., MS rt→lf	infarction of the right hemisphere
10	55/m	03	1	8 hrs.	acute SDH rt., MS rt→lf cont.front.lat.rt	polytrauma
11	63/m	03	1		acute SDH lf., MS lf→rt cerebellar haemorrhage	
12	71/m	07	1	1,5 hrs.	rt. temp.cont.	
13	75/m	03	1	1 hr.	SDH lf, MS lf→rt rt. temp. cont.	occ.infarction lf

Table 13. *Brain stem haemorrhages group C*

No	Age/Sex	GCS	GOS	Interval since accident	CT-Findings	Remarks
1	14/m	04	1	1,5 hrs.	cont.fr.lf>rt. subependymal haemorrhage lf ventricle	
2	16/m	04	2	2 hrs.	thalamic haemorrhage rt.	
3	18/m	05	3	3 hrs.	rt.fr.dorsal haem.cont. subependymal haemorrhage rt. ventricle	
4	19/m	06	4	2 hrs	lf. temp. cont.	
5	18/m	05	1	3 hrs	lf. temp. and fr.dorsal cont,pinpoint, thalamic haemorrhage rt. subependym. haemorrhage rt.ventricle	diabetes insipidus
6	21/m	06	1	2 hrs.	compound fracture fr.lf. IVH	open head injury anterior infarction bl. diabetes insipidus
7	22/m	04	1	2 hrs.	acute SDH rt,MS rt→lf. IVH	
8	26/m	07	2	3 hrs.	EDH. lf.fr,impr.fracture rt.fr., IVH	
9	32/m	04	4	5 hrs.	compound fracture rt. temp. rt. temp.cont.	
10	39/m	06	3	2 hrs.	cont. temp. lf.	open head injury
11	48/m	04	1	3 hrs.	acute SDH bl.mult.cont.bl.	open head injury
12	51/m	06	2	1 hrs.	subependymal haemorrhage lf. ventricle , IVH	polytrauma
13	63/m	09	1	2 hrs,	both sided chronic SDH	thrombosis of sag.sinus intraop. (histol.proven)

stem haemorrhage. By correlating the mass shift with the chosen classification into the group A, B and C the effect of the associated injuries and the mass shift caused by them is clearly shown (Table 16).

Table 14. *Density of brain stem haemorrhage*

group	N	HU 20-29	30-39	40-49	50-59	60-69
A	6	-	1	2	3	-
B	13	1	5/4	1	2	-
C	13	-	4/1	3/1	2/1	1

Table 15. *Patients with brain stem haemorrhage.* Correlation between findings on admission, mass shifts and outcome

mass shift	GCS	GOS N	1	2	3	4	5
absent	03-05	11	6	1	2	2	-
	≥ 06	9	3	1	-	3	2
less 1 cm	03-05	2	2				
	≥ 06	-					
less 2 cm	03-05	4	4				
	≥ 06	2	1	1			
more than 2 cm	03-05	4	4				
	≥ 06	-					

Table 16. *Mass shifts in patients with brain stem haemorrhage*

Group	N	mass shift absent	less 1cm	less 2cm	more than 2 cm
A	6	6	-	-	-
B	13	3	2	5	3
C	13	11	-	1	1

Clinical Findings

Only two patients were conscious at the time of admission, whereas 21 (66%) were in a decerebrate state (GCS 03–05, Table 17). No disturbance of the size

Table 17. *Patients with brain stem haemorrhage.* Correlation between findings on admission and outcome in groups A–C

Group	GCS	N	GOS 1	2	3	4	5
A	03-05	3	1	-	2	-	-
	≥ 06	3	-	-	-	1	2
B	03-05	11	11	-	-	-	-
	≥ 06	2	2	-	-	-	-
C	03-05	7	4	1	1	1	-
	≥ 06	6	2	2	1	1	-

of the pupils and the light reaction were detected in group A. They were present in 10 of 13 cases in group B and in nine out of 13 cases in group C (Table 18). The bilateral light reaction showed the same proportions. Alltogether 20 of 32 patients died, which included only

one in group A from an extracerebral cause and all of group B.

Table 18. *Patients with brain stem haemorrhage.* Correlation of pupil size, shape, light reaction and corneal reflex

Group	N	pupils equal	diff.	dil.	lightreaction equal	diff.	neg.	cornealreflex equal	diff.	neg.
A	6	6	-	-	6	-	-	5	1	-
B	13	3	3	7	3	-	10	3	-	10
C	13	4	8	1	4	5	4	8	4	1

Discussion of Traumatic Brain Stem Haemorrhages

Haemorrhages with a smaller volume that 0.1 ml can not yet be detected in the CT[22], so that only the larger ones can be seen *intra vitam*, and the statements about their frequency still remain uncertain. Jacobs[47] in a comparative CT and morbid anatomical study found 50% incorrect diagnoses with involvement of the brain stem. George[36] in a correlation of CT and neurological findings saw 15% primary brain stem lesions, without however mentioning the frequency of haemorrhages. The first systematic study presented here provides a clue, with a figure of 4%.

Corresponding to the neurological definition and findings, secondary brain stem haemorrhages are always associated with a herniation. On the other hand, according to Meyer[78] the site of predilection for primary brain stem haemorrhages caused by rhexis is, with decreasing frequency, the midbrain lateral sulcus opposite the tentorial margin, the tegmentum and ventrally the irrigation area of the intercrural branches of the basilar artery.

Patients with such bleeding from ruptured arteries or veins have not survived at all, or only for a few hours. We were able to confirm in the CT the sites of these lesions. A haematoma in the floor of the fourth ventricle was one unusual observation. Whether in this case of suboccipital trauma it was the result of a central cavitation effect[114] must remain open to question. The location is specified in a schema by Courville (quoted by Krauland[62]) as the site of predilection and is also mentioned by Lindenberg[72, 73].

In addition the site of the secondary haemorrhage of group B corresponds to those described by Meyer[78], predominantly in the midline, extending from the midbrain down to the lower third of the pons.

In group C the haemorrhages opposite the tentorial margin must be regarded in view of their location, as primary rhexis bleeding. In two patients it was not possible to confirm this by autopsy. Indeed, because of the increased survival time[130], the differentiation between primary and secondary lesions is becoming more difficult, even histo-pathologically. One patient was particularly interesting: narrowed cisterns in the first CT gave no definite support for the proven haemorrhage which was apparent in the second CT. The site was typical for a secondary haemorrhage. However, it is also conceivably a haemorrhage in a primary focus of contusion, as is recognised in the case of trauma in other parts of the brain[64].

In the CT in two patients the uncus was hyperdense on both sides in its posterior portion. According to Lindenberg[72], this is a site for primary contre coup lesions.

The survival rate, of 37.5% for all brain stem haemorrhages is significantly better than would be expected according to the neuropathological findings[49, 78, 79, 95, 96].

Isolated examples are well-known, such as two observations with several years survival[46, 56a]. The CT allows for the first time the recognition and supervision of larger haemorrhages and thus, an assessment of the prognosis. Of the direct primary traumatic haemorrhages five of six survived, of the secondary type non, and seven of thirteen of the mixed type. The extent of the herniation is of the greatest significance for the prognosis, in the secondary and mixed forms. Our findings confirmed those of Tsai[128].

In the differentiation of the types of haemorrhage, the temporal behaviour, location, findings in the cisterns and the density, provide the important clues. Primary haemorrhages develop more frequently in the young. Older patients more frequently show secondary haemorrhages in the same way that intracerebral traumatic mass haemorrhages are also more frequent in them. Here also, the multifactorial pathogenesis is of increasing importance.

Summary

Traumatic haemorrhages in the brain stem develop in 4% of all severe cranio-cerebral injuries. Isolated primary (20%) and pure secondary ones (40%) can be distinguished on the basis of their location, demarcation, density and the extent of the herniation. Primary haemorrhages without signs of herniation were survived, but patients with secondary haemorrhages and with completely obliterated cisterns all perished. Mixed clinical entities require individual analysis. Their prognosis is essentially determined by the associated injuries and the mass shifts which they

cause. As with the other parts of the CNS, secondary haemorrhages seem to be met within foci of primary contusion.

3.1.2. Spontaneous Brain Stem Haemorrhage

After a first description of a spontaneous pontine haemorrhage by Cheyne 1812 (cited by[28]), Gowers 1893 and Oppenheim 1900 described the symptoms. The frequency was reported between 6%[28], 9%[33], and 9.9%[50], in which all ages over 30 years are represented, with the main emphasis between 40 and 60 years. As regards aetiology hypertension[28] or else hyalinosis of the arterioles is mentioned in 97%. Vascular malformations[25, 125] or the "cryptic arteriovenous malformation" are the second important cause[23]. The most frequent location is pontine, unilateral, bilateral or segmental[115]. Only exceptionally the haemorrhages were located in the midbrain[111]. The first successful operation on a pontine haematoma was reported by Dandy in 1932. Reports of single cases followed after 1960[8, 54, 60, 89]. With the introduction of computed tomography, reports on brain stem haematomas became more frequent[27, 93, 94] as well as successful operations on midbrain[32] and pontine haematomas[29, 92, 122].

Sano[105, 106] reported on probably the largest series of recent years. He classified the haematomas for the first time on the basis of their size and extent, and correlated this with the clinical course and prognosis.

Material

Our own material includes 22 spontaneous haemorrhages, 18 men and four women; secondary haemorrhages from supratentorial or infratentorial tumours have been excluded. Hypertension in the fifth to the seventh decades is prominent in the aetiology. Three patients had a vascular malformation, two a haemorrhage into a tumour and in the case of the two patients who survived, the aetiology remains abscure. Three patients were younger than 40 years, the youngest at 26 years survived a typical localised hypertensive haemorrhage in the right lamina tecti. With the exception of one patient all those over 50 years had a hypertensive haemorrhage, frequently associated with diabetes mellitus. Of the 15 with hypertensive haemorrhages two, which were located atypically in the midbrain, survived.

Results

CT Findings

Fifteen hypertensive haemorrhages, four mainly in the midbrain and eleven pontine were located by computer tomography. Using the classification of Sano[106], two of the eleven pontine haematomas extended over one third of the cross-section of the brain stem, the reminder involved two-thirds and more. However, even the two smaller haematomas also crossed the midline. The large haematomas were situated nine times in the midline and twice on the right side. In all cases they were very close to the fourth ventricle.

The cisterns were locally obliterated and in the case of the pontine lesion were narrow and obliterated below the tentorium, but were visible above and below the haematoma. Ten haematomas extended in an axial direction for more than 3 cm and the density was between 30 and 60 HU (Table 19). Eight patients developed a hydrocephalus, in six the cisterns were locally, and in two completely, obliterated. Of fourteen patients who showed no hydrocephalus in the first CT, nine died within 24 hours, without a further CT beding done.

Table 19. *Density of spontaneous brain stem haemorrhages*

Density HU	30-40	40-50	50-60	60
Survivors	2	2	-	-
Dead	1	6	8	3

Clinical Findings

Of 22 pts, 13 were in a decerebrate state on admission (Table 20). Four of the remaining nine survived. Two of

Table 20. *Findings on admission and mortality of spontaneous brain stem haemorrhage*

GCS	Dead	Survived	N
03-05	13	-	13
06-08	4	2	6
>08	1	2	3
	18	4	22

these had atypically situated midbrain hypertensive haematomas and two had haematomas of undetermined aetiology with a density of 30–40 HU. The clinical symptoms on admission were determined by disturbance of consciousness (N = 20), lateralised sensori-motor symptoms (N = 14), disturbances of eye movements (N = 11) and of respiration (N = 5). Abnormalities of the pupils were found in 15, of whom five had anisocoria, and in six the corneal reflex was impaired. The density of spontaneous hypertensive brain stem haematomas was between 40 and 60 HU (Table 19).

Discussion of Spontaneous Brain Stem Haemorrhages

Aetiology and age distribution as well as the mortality correspond with the figures well-known from the literature[28, 106, 115]. In those patients who survived, the neurological symptoms correlated with the CT findings. The obliteration of the cisterns takes place only locally, but little or no significance can be given to the herniation from the CT point of view. The cross-section of the brain stem is locally enlarged. The density measurements submitted here, although not significant, offer new points of view regarding these lesions. Hypertensive spontaneous brain stem haematomas show density values which correspond to the primary traumatic haemorrhages. With all due caution therefore, only limited significance can be given to the density measurement in relation to the differential diagnosis, as spontaneous haematomas with a density below 50 HU may have a cause other than hypertension.

Summary

Computer tomography has made the diagnosis of spontaneous brain stem haematomas more precise. Because of the very high mortality the aetiology, frequency, age distribution and the site of the lesions, are well-known from post mortem studies. When considering the extent of the bleeding in the brain stem, the CT findings are certainly of some prognostic significance. The obliteration of the cisterns is locally limited and only of significance for the development of hydrocephalus. The degree of impairment of consciousness at the time of admission appears to be the best prognostic criterion.

3.2. Hypodense Brain Stem Lesions

Compared with the acute hyperdense lesions which are definitely pathognomonic of a brain stem haematoma, the hypodense lesions of the brain stem do not allow any definite pathogenetic classification. Softening, inflammation, tumour and oedema all have to be considered. As with acute hyperdense haematomas of the brain stem, clues to the differential diagnosis are given by the distinction between acute direct hypodense injuries without associated cistern involvement and indirect lesions with cisternal hernias and an enlarged or normal brain stem.

3.2.1. Brain Stem Infarcts

Out of 135 cerebral infarcts seen in the neurological and neurosurgical clinics in Giessen[15] (Table 21) 28 (21%) involved the brain stem. There were positive findings in the CT in only seven cases (25%), five were doubtful

Table 21. *Frequency of positive, doubtful and negative CT findings, comparing supratentorial, brain stem and cerebellar infarcts*

Location	CT findings					
	positive		questionable		negative	
supratentorial N= 97	82	84,5%	–	–	15	15,5%
brainstem N= 28	7	25 %	5	18%	16	57 %
cerebellum N= 10	10	100%	–	–	–	–

and sixteen were negative. On the other hand, ten cerebellar infarcts (7%) were always, and 82 of 97 supratentorial infarcts (almost 85%) were apparent in the CT.

Material

Seven patients (four men and three women) with a brain stem infarct confirmed in the CT were between 40 and 80 years old. As risk factors there was a vascular hypertension in four, diabetes mellitus in one and in one patient both conditions. In one patient the infarct was traumatic in origin as a result of a brief strangulation.

CT Findings

In the CT the areas of infarction could be detected on two occasions within one week, three times within two, and once within four weeks. In the strangled patient it was detected immediately.

Positive CT findings are present solely in the pons and in the ponto-mesencephalic junction. Infarcts in the medulla oblongata could not be identified, although three patients presented a Wallenberg's syndrome. In all seven patients the cisterns were open and there was never any swelling or compression of the brain stem. With contrast medium none of the cases showed any enhancement or any other evidence of disturbance of the blood-brain barrier.

The density of the infarcted area was on average 15.3 HU, with variations between 5.5 and 23.3 HU. The aquaeduct was always patent, with no evidence of hydrocephalus.

Clinical Findings

Six patients were somnolent on admission, one patient with bilateral brain stem infarcts was comatose and the patient who had been strangled was decerebrate. The previous history showed that one patient had already suffered a supratentorial infarct and another a coronary infarction. The presenting and most frequent symptoms were giddiness, nausea, headache, unilateral sensori-motor disturbances and unsteadiness of gait.

All patients presented a dysarthria or even an anarthria, facial paresis and dysphagia. Two patients presented with a hemi-ataxia, one in association with a paresis of horizontal ocular movement. One patient presented with a picture of bulbar paralysis. They all survived with severe neurological deficits and the primary comatose patient remained apallic.

Discussion of Brain Stem Infarcts

A CT correlation was only found in 25% of the clinically verified brain stem infarcts. This corresponds with the observations of Busse[15] and Feistner[34]. In the supratentorial space they were successful demonstrated in 85% and thus, more frequently than indicated by Sager[102]. The more up-to-date CT apparatus may be the explanation for this. On the other hand brain stem infarcts are usually small and occur in the area for maximal artefacts and where the CT has minimal resolution ability. Even special recording techniques (high lightening, overlapping slices, comparative density measurements) do not increase the proportion that can be identified in the CT. There was no contrast enhancement in any case. Infarct oedema is very poorly marked and we never saw any enlargement of the brain stem, corresponding to which the cisterns were always open and there was no hydrocephalus.

Summary

The CT diagnosis of brain stem infarcts is unsatisfactory. Only one in four can be verified by the CT. The main reasons for this are the minimal surface extent and depth, as well as the absence of any associated manifestations such as distension of the brain stem, hernias and contrast enhancement. Technical factors play an additional negative role. Thus, infarcts of the medulla cannot be identified. The clinical state and neurophysiological findings are decisive for the diagnosis.

3.2.2. Basilar Artery Occlusion

Six pts with a verified basilar artery occlusion presented an atypical pattern. The transversely extending hypodense lesion was always recognisable in the CT.

Material

The six pts with basilar occlusion were aged between 24 and 62 years. Five were male and the youngest was female. In her case as risk factors, there were a year-long continuous administration of oral contraceptives and also excessive smoking. One patient survived for 14 days after open CSF drainage and another, eight days with a "locked-in" syndrome.

CT Findings

The CT findings in our cases were uniform. The transversely extending hypodensity of the brain stem is apparent in all slices. In two cases both cerebellar hemispheres, and three times the left only were also infarcted. In all cases the cross-section of the brain stem was enlarged and the aqueduct obliterated. In the case of an associated unilateral cerebellar infarct the brain stem was also correspondingly compressed and the aquaeduct displaced to the opposite side and obstructed. In five cases all the cisterns were obliterated. An exception was the patient with the "locked-in" syndrome. In this case the cisterns were narrow above the hypodense region.

The average *density* of the infarct in the first CT was 18.2 HU with variations between 12.4 and 22.5 HU. The density in the surrounding tissue was always at least 8 HU higher than the infarcted area. Contrast medium was not given at all. One patient survived for 14 days after a drainage operation to relieve pressure. After very marked hypodensity on the third day (9.5 HU) he showed an isodensity on the eleventh day (24.1 HU).

Hydrocephalus developed in all patients with basilar occlusion and associated cerebellar infarct. One patient

survived for 14 days after the insertion of a CSF drain. The ICP before the relief of pressure was 50 mm Hg, with a maximum reading of 60 mm Hg and the appearance of plateau waves.

The *clinical findings* are set out in Table 22. After a prodromal stage of 1–3 days with marked pains in the head and neck, they showed giddiness and disturbances of consciousness going on to a decerebration or bulbar syndrome. In patient 5 an intermediate stage of "locked-in" syndrome occurred. Angiographically a basilar artery occlusion between AICA and posterior cerebral artery could be demonstrated. Figure 8 shows the CT finding. All patients died.

Discussion of Basilar Artery Occlusions

In the cases of embolic or thrombotic basilar artery occlusion and depending on the collateral circulation, infarcts of the medulla oblongata, the pons and the midbrain develop. Anastomoses through the cerebellar plexus connect up the vascular territory of the superior and the posterior inferior cerebellar arteries[31]. The duration of the prodromi can be explained by the progress of the thrombotic occlusion and, with that, the inadequately developing anastomoses. The involvement of the tegmentum in the CT explains the coma[31, 116] with occlusion of the rostral portion of the basilar artery. Silverstein surveyed 83 patients with pontine infarcts and only two of these pts survived 2–3 months. The mortality agrees with our observations.

Table 22. *Clinical findings in six patients with basilar infarction*

Patient	1	2	3	4	5	6
Age	24	34	43	55	61	62
Sex	f	m	m	m	m	m
Prodromi (days)	2	1	3	2	1,5	1
GCS	03	04	03	04	04	04
Pupils R/L	dil./dil.	small/small	R < L	R > L	small/small	small/small
Light reaction R/L	- / -	- / -	- / -	- / -	- / -	(+) / (+)
Corneal-reflex R/L	- / -	(+)/ -	- / -	-/ (+)	- / -	+ / +
Tone	diminished	elevated	diminished	elevated	elevated	diminished
Resp'n	insuff.	spontan	insuff.	insuff.	spontan	insuff.
Reaction to pain	no	dec.rig.	no	dec.rig.	dec. rig.	dec.rig.
Density	19,3	14,3 9,5 24,1	16,2	18,7	12,4	22,5
GOS	1	1	1	1	1	1

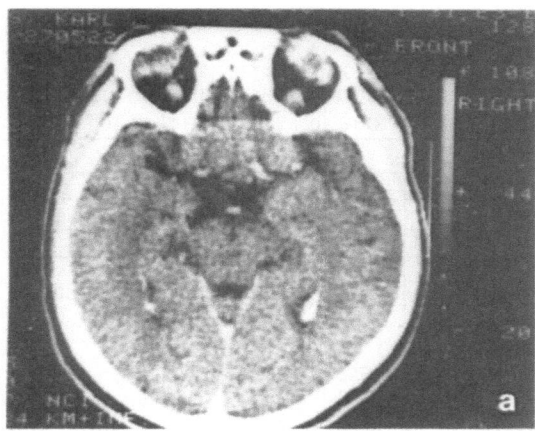 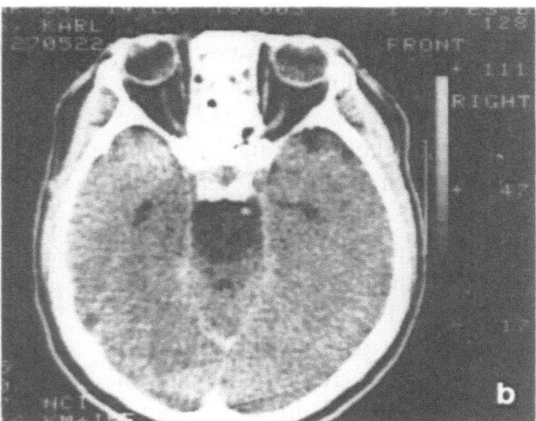

Fig. 8. CT findings in a locked-in-syndrome

In 1982 Zeumer[136] reported on successful fibrinolytic treatment of a basilar thrombosis in a 27 years old woman. With more experience in a larger number of pts the results were much less favourable[26]. Kollikowski[59] also reached the same conclusion that the prognosis of basilar thrombosis in older subjects with other risk factors must be regarded as unfavourable. In our cases intraarterial fibrinolysistherapy was not used.

In patient 5 it was possible to demonstrate in the CT an intermediate stage with clinical correlation before the onset of infarction of the midbrain. The patient survived this incident seven days. It is infarction of the cerebellum which determines the extent of the cisternal obliteration. In the case of isolated infarction of the brain stem, with a swollen brain stem, the cisterns are narrowed. As regards the basic illness the obliteration of the cisterns is, however, only of significance for the development of hydrocephalus.

Summary

In primary vascular episodes the transversely extending hypodensity of the brain stem in the CT is pathognomic of a basilar artery occlusion. The cerebellum is selectively involved, and in these cases the cisterns are obliterated and hydrocephalus develops. When the cerebellum is not involved the cisterns are narrowed and there is no hydrocephalus. The reduction of the density is variable and amounts to at least 8 HU compared with the uninfarcted area. The prognosis in most cases is unfavourable.

4. Indirect Secondary Infarcts of the Brain Stem and Other Regions

Secondary infarcts of the brain stem and other parts of the brain are found in acute, primary, unilateral and bilateral traumatic and spontaneous lesions. They are the sequelae of an acute rise in intracranial pressure with accompanying mass shifts and cisternal hernias. We are not dealing here with infarcts in arteriosclerotic and embolic vascular lesions, nor with arterial spasms, which may occur after aneurysmal subarachnoid haemorrhage and the possible resulting hypoxic and anoxic damage.

Survey of Literature

Infarcts of the occipital lobe occur most frequently in space-occupying lesions, and since the first description by Meyer in 1920[81] have been repeatedly investigated[83, 101, 109]. Their frequency was estimated to be 3%[109]. The causes are predominantly tumours and subdural haematomas. The first systematic investigation (Pia, 1957) was based on 26 cases in the literature and eight personal cases (material Hallervorden, Spatz, and Tönnis).

Acute infarcts occurring with cranio-cerebral injuries play a subordinate role in clinical investigations. Detailed post mortem studies began with Lindenberg[71, 72] who emphasized in the aetiology the importance of hypoxaemia and acidosis resulting from inadequate blood flow and the subsequent oedema. He maintained that it was uncommon to see tissue destruction on its own as a result of impaired venous drainage.

According to Kleihues[57] infarcts of the posterior cerebral artery from intracranial mass shifts are, pathologically, haemorrhagic infarcts. These, from their situation in the territory of the posterior cerebral artery, are extremely variable. They are according to his opinion, arterial in origin. Jellinger and Seitelberger[49] observed in their material (576 autopsies on head injuries, who had survived for more than ten days) only seven examples of ischaemic necrosis of the occipital lobe and they saw no infarcts in other vascular territories which Lindenberg[71] mentioned, apart from the focal lesions described. Adams[1] examined the brains from head injuries in whom the intracranial pressure had been continuously monitored during life and found inter alia as a correlate of the rise of intracranial pressure, infarction of the medial occipital cortex, which he interpreted as a sequel of compression of the posterior cerebral artery. He also described the ischaemic necrosis which was particularly frequently found in the border zones of the territory of the anterior and middle cerebral arteries. Zülch[137] mentioned that in rare cases other arteries than the posterior cerebral could be compressed, as for example the anterior cerebral artery.

Apart from the focal ischaemic lesions, Graham and Adams[37] first described the border zones which they were able to identify in 91% of 151 cranio-cerebral injuries, and also lesions which involved the entire arterial territory. In their material this twice involved the anterior cerebral artery (once bilaterally), seven times the middle cerebral artery (twice bilaterally), five times the anterior and middle cerebral arteries bilaterally, as well as the posterior cerebral artery thirty times, including 14 which were bilateral.

It followed from these findings that ischaemic infarcts were the result of disturbances of the arterial circulation. They occur with extreme rises of intracranial pressure, up to complete arrest of the intracranial arterial circulation. The cisternal hernias and, in certain cases, direct vascular lacerations are responsible for the particular location of these lesions.

In contrast to these extensive morphological and clinical reviews, there are no systematic CT investigations and comparable clinical observations.

Material

Out of 253 pts 37 (14.6%) showed infarcts of an arterial territory. For spontaneous haemorrhages this corresponded to a frequency of 53 : 3 or 5.6% and for cranio-cerebral injuries 174 : 32 or 18.4%. No particular age incidence was noted. The occipital lobe, involved on 20 occasions, was the most frequently affected (Table 23) and included 15 times unilateral ipsilateral, once contralateral to the space-occupying lesion, and four times bilateral.

In four cases the rostral brain stem and the basal ganglia were also involved. Hemisphere infarcts were found unilaterally, twice without and three times with brain stem involvement. The temporo-parietal lobe was once, and the frontal lobe twice, infarcted bilaterally, once including the rostral brain stem. Pts who survived for a longer period were found, with only one exception, in the group of posterior artery infarcts.

Table 23. *Site of secondary infarcts in the case of acutely raised intracranial pressure*

Site	N	Lateralisation ipsi-	contra-	bil.	Brain stem involvement	Additional infarcted areas
Frontal	2			2	1	
Temporal	1			1		
Occipital	20	15	1	4	4	Basal ganglia 4
Hemispheric	5	5			3	
Brainstem isolated	9				9	Hypothalamus 1

CT Findings

The correlation between infarcts, mass shifts and cistern findings on the first day, at the time of establishing the diagnosis of the infarct and the survival time after this is shown in Table 24.

Only ten pts showed no lateralised mass shift, of whom four had bilateral infarcts and in these the transtentorial herniation was corresponding strongly marked. In the first CT the cisterns were completely obliterated 31 times and narrowed six times. In this connection the global assessment "cisterns obliterated" signified that none of the basal cisterns could be identified any more. When the cisterns were reported "narrowed" the cisterna quadrigemina and the cisterna Galeni were each still visible on three occasions and the cisterna Galeni alone on two, while all other cisterns were obliterated.

At the time of identifying the infarct the cisterns were obliterated 26 times and were narrowed eleven times, including two pts with narrowing of only the basal cistern and only the cisterna quadrigemina. The cisterns were not free i.e. without herniation in any cases. In ten pts the hypodensity was already recognizable on the first day of the incident, and among these the brain stem alone was involved eight times. Of the remaining infarcts eleven were recognisable on the first, five on the second and four each on the third and fourth days. As the CT scans were not made daily it is quite possible that infarcts appeared even earlier. In three patients the infarct appeared on the 6th, 12th and 14th days respectively.

Haemorrhagic infarcts in a CT without contrast medium, recognised by zones of increased and decreased density alongside each other were seen four times, on two occasions immediately, once after nineteen hours and one four days after the identification of the infarct. Contrast medium was given eight times with a CT recording, twice there was no enhancement or any extravasation of contrast. On three occasions the extravasation of contrast could be seen in

Table 24. *Correlation of cistern findings, mass shifts and survival time in relation to the passage of time in secondary infarcts*

Vascular territory	N	Shift 0	-1	-2	>2	cisterns first CT obl.	nar.	cisterns when infarct diagn. obl.	nar.	time of diagnosis of infarct 0	1	2	3	4	>4	survival time after diagn. of infarct <1	2	3	4	>4
Ant.cer.art.bil.	1	1				1		1				1				1				
bil.brainst.invol.	1	1				1		1			1						1			
Middle cer.art.bil. without brainstem	1	1				1							14d							13d
Hemisphere unilat. without br.stem	2		2			1	1	2		1	1							1	1	
unilat.br.st.invol.	3		3			3		3			2	1				1	1	2		5d 9d
Post. cer. without brain st.ipsilat.	13		3	8	2	12	1	6	7	5	2	3	3			5	1			2 S. 15 d 22 d 27 d
contralat.	1		1				1	1					1			1				
unilat.br.st.and bas. gangl.involved	2			1	1			1	1	1	1						1			6 d
bilateral	2	1		1			2	1	1	1			12 d							18 d 1 s.
bilat. br.st.and bas. gangl.involved	2	2				2		2		2						2				
Brain stem isolated	8	5		3		7	1	6	2	7				6 d		3	2		2	1 S.
Br.st. and hypoth.	1	1				1		1		1						1				
	37	10	6	18	3	31	6	26	11	10	11	5	4	4	3	13	8	2	2	4 S.

s = survivors

the first CT and once each after 36 hours and after four days.

Those pts who survived longer, were found, with only one exception in the group of the posterior cerebral artery infarcts. The hypodensity in the posterior territory regresses with variable speed. In 14 pts, after the entire territory was affected, three times the end stage could be detected as merely an enlargement of the subarachnoid space alongside the falx, over the medial aspect of the occipital lobe. All the other infarcted areas were variable in size and affected principally the occipital lobe in its basal part and medially where it is situated alongside the falx.

In the case of fresh infarcts the difference in *density* between the infarcted tissue and the rest of the brain tissue is 4–18 HU with a clustering between 7 and 11 HU. In eight cases the density could be observed in relation to the time of observation. Three times an isodensity was detected between the seventh and the twelfth day. In the late stages the difference in density was, at 18–22 HU, distinctly higher than in the acute stage.

Out of 37 pts, 13 showed a *hydrocephalic dilatation* of the opposite lateral ventricle. In all cases the midline shift was more than one cm. One case with an epidural haematoma in the posterior fossa developed hydrocephalus. In view of the severity of the injuries in this group of pts, the development of a hydrocephalus did not appear to have any effect on the course of the disease.

Angiographic Findings

In six of 37 pts an angiogram was done after the occurence of the infarct. In those pts with an extensive cranio-cerebral injury who showed an infarct in the territory of both anterior cerebral arteries, including also the brain stem, only the A 1 segment of the anterior cerebrals bilaterally and the anterior communicating artery were shown. In the patient with a hypodensity in the brain stem, including the hypothalamus, there was no filling of the vessels in the posterior fossa. At the autopsy the whole of the brain tissue in the posterior fossa was liquified and could not be processed histologically. All other angiograms involved pts with infarcts in the area of the posterior cerebral artery circulation. It was not possible to detect any interruption or compression of the arteries in any of these pts.

Clinical Findings

Twenty-three of these 37 pts were admitted in a decerebrate state and non of them survived (Table 25). Of the remaining 14 pts 4 survived. Two of these had a supratentorial epidural haematoma and one each had a

Table 25. *Ratio GCS/GOS in patients with secondary infarcts*

GCS	N	GOS 1	2	3	4	5
03-05	23	23	-	-	-	-
≥06	14	10	-	1	2	1

Table 26. *Neurological findings in patients with secondary infarcts, with and without brain stem hypodensity*

	N	Pupils		Light react.		Corneal reflex		Respiration		Tone		Brainstem Haemorrhage	Diabetes insipidus	GOS	
		dil.	rest.	absent	pos.	absent	pos.	absent	pos.	diminish.	normal			1	2-5
Hypodensity	17	11	6	14	3	11	6	10	7	9	8	8	3	16	1
Isodensity	20	2	18	6	14	6	14	1	19	1	19	6	1	17	3
Significance		p < 0,01		p<0,01		p< 0,05		p < 0,01		p< 0,01		–	–	–	

supratentorial subdural and an infratentorial epidural haematoma. Pupillary reactions, light reflexes and corneal reflexes, muscle tone and respiration were variably affected. By distinguishing the groups with and without brain stem hypodensity (Table 26) it became apparent, that hypodensity of the brain stem had a significant relationship to dilated, inactive pupils, absent corneal reflexes, absent spontaneous respiration and loss of tone.

The large number of associated secondary brain stem haemorrhages (N = 14), as shown in Table 26, underlines the great variety of secondary vascular damage resulting from the acute rise of intracranial pressure. At the same time they explain the uniformity of the clinical syndrome, the difficulty of a topical classification and the unfavourable prognosis. Four cases with diabetes insipidus, two associated with anterior infarcts, once each in a hemisphere infarct and in a brain stem infarct, indicate the involvement of the hypothalamus. A CT diagnosis is not possible.

One patient with a hypodensity of the brain stem visible in the CT survived without any neurological deficits. He was drowsy on admission and in the CT showed, apart from a herniation through the foramen magnum, an ascending herniation from an epidural haematoma in the posterior fossa. This unusual finding cannot be explained from the CT appearances alone.

These secondary infarcts were survived by 57% of the pts for only two days and 73% died within a week. In all only 11% of the pts survived.

Discussion of the Indirect Secondary Infarcts

Infarcts in cases of intracranial space-occupying lesions, mostly head injuries, can be recognised in the CT if they involve individual vascular territories[52]. We found their frequency was 14.6% or for head injuries alone was 18.4%. Graham[37] found them in 20%. Smaller infarcts in the border zones, which were described in detail from post mortem material[2, 37, 49, 95, 96, 97], could not be differentiated from areas of traumatic oedema.

Infarcts which are arteriosclerotic in origin are hypodense in 75% and isodense in 25%[102]. Three stages may be distinguished:

Stage 1: First week, fresh infarct, characteristically vague outline and no contrast enhancement.

Stage 2: Second to fourth week, subsiding of the infarct oedema, frequently marked contrast enhancement. The granulation tissue, which is formed as a result of the phagocytosis is regarded as responsible for this contrast enhancement.

Stage 3: After the fourth week porencephalic cyst.

Patients with infarcts who show a contrast medium enhancement seem to have a much worse prognosis than the rest[100]. Infarcts can display a so-called "fogging effect" i.e. they can become isodense[13, 112] in the second to third weeks.

The clinical symtpoms which are determined by the infarcts are not those of the primary disease. Actually the infarcts are discovered in the CT rather fortuitously and are associated with a further deterioration in the pts clinical condition.

When the densities are reduced to 4–18 HU, with a clustering between 7 and 11 HU, it admits of no differentiation from reduced densities from other causes such as oedema from infarcts, trauma or tumours[53].

The anatomical pattern in the CT indicates which vessels are infarcted. In 14 of 20 infarcts of the occipital lobe the whole area of distribution of the posterior cerebral artery was involved and in its further course regressed to a marked degree, so that in the CT there was only evidence of dilatation of the subarachnoid spaces over the occipital lobe alongside the falx. It seemed unlikely that it involved the formation of an isodensity in the region of the infarcts[13, 112]. It is more likely that it had been a question of a temporary circulatory disturbance of arterial origin with subsequent resolution of the infarct oedema.

In only 6 of 20 cases the hypodensity of the occipital lobe did not correspond to the entire territory of the posterior cerebral. It was seen once in the first CT, three times on the first day and once each on the fourth and the twelfth day after the incident. The possibility exists, that the entire territory is already involved beforehand

and the CT was done at a possible stage of regression. In these cases a disturbance of venous drainage might be considered, whereby the CT only once depicted a haemorrhagic infarct.

Sato[107] was able to show in animal experiments, that venous congestion produces a destruction of the blood-brain barrier. Sano[104] in animal experiments in 1982 established that in the early stages of raised intracranial pressure there is a venous stasis in the presence of slight changes of the CBF. At a later stage of raised pressure, the venous blood was "squeezed out" by the raised intracranial pressure. At the same time, however, with a pressure of 50 mm Hg an arterial stasis with subsequent oedema developed.

The prognosis for patients with infarcts, in the case of acute raised intracranial pressure, is unfavourable, 73% dying within one week of the demonstration of the infarct and only four patients surviving (11%). Patients who survived for more than one week, were found with one exception (Table 24) in the group of occipital lobe infarcts. This must be the explanation why Jellinger and Seitelberger[49], who investigated a group who had survived at least ten days, observed only posterior infarcts and no infarcts in any other arterial territory.

Summary

Indirect secondary infarcts of the brain stem and other areas develop in 15% of cases with acute intracranial lesions with raised intracranial pressure. Herniation into the tentorial hiatus is marked in severe lateralised mass shifts (73%) or in more severe transtentorial herniation. In 92% the infarct develops within the first four days after the incident.

Only 27% of the pts survived this additional space-occupying lesion for longer than one week. Infarcts in the posterior cerebral artery territory can be more easily compensated for, and survived. A transversely extending hypodensity of the brain stem is nearly always an unfavourable sign.

Acknowledgement

I am grateful to Mr. Charles Langmaid for translating this paper.

References

1. Adams JH (1975) The neuropathology of head injuries. In: Vinken PJ et al (eds) Handbook of clinical neurology, vol 23. North Holland Publ Comp, Amsterdam New York, pp 35–65
2. Adams JH, Mitchel DE et al (1977) Diffuse brain damage of immediate impact type. Brain 100: 489–502
3. Agnoli AL, Cristante L et al (1980) Brain herniation by cranial computer tomography: clinical radiological correlations. In: Anatomy-physiology in CT. Kugler Med Publ, Amstelveen
4. Agnoli AL, Laun A et al (1980) Computerized tomography in brain stem haemorrhage. Diagnostic and prognostic aspects. Lecture IXth Congress of European Society of Neuroradiology, Brussels
5. Agnoli AL, Busse O et al (1980) Prognostic significance of subarachnoidal CSF spaces abnormalities in CT. Lecture IXth Congress of European Society of Neuroradiology, Brussels
6. Ambrose J (1973) Computerized transverse axial scanning (tomography). Part 2: Clinical application. Br J Radiol 46: 1023–1047
7. Ambrose J (1974) Computerized X-ray scanning of the brain. J Neurosurg 40: 679–695
8. Arseni C, Stanciu M (1973) Primary haematomas of the brain stem. Acta Neurochir (Wien) 28: 323–330
9. Auer LM et al (1980) Relevance of CT-scan for the level of ICP in patients with severe head injury. In: Shulman K et al (eds) Intracranial pressure IV. Springer, Berlin Heidelberg New York, pp 45–47
10. Auer LM, Auer T, Sayana J (1986) Indications for surgical treatment of cerebellar haemorrhage and infarction. Acta Neurochir (Wien) 79: 74–79
11. Bagley C (1932) Spontaneous cerebral haemorrhage. Arch Neurol Psychiat (Chic) 27: 1133–1179
12. Bauer BL (1969/70) Protein and amino acid metabolism in central dysregulation. In: Modern aspects of neurosurgery, Proc Ann Meetings. Deutsche Gesellschaft für Neurochirurgie, vol 1, ICS 242. Excerpta Medica, Amsterdam 1971 a
13. Becker H, Desch H et al (1979) CT fogging effect with ischemic cerebral infarcts. Neuroradiol 18: 185–192
14. Berger MS, Pitts LH et al (1985) Outcome from severe head injury in children and adolescents. J Neurosurg 62: 194–199
15. Busse O, Agnoli AL, Feistner H (1980) Beziehungen zwischen neurologischen Ausfällen und computertomographischen (CT) Befunden bei Hirnstamminfarkten. In: Verh Dt Ges Neurol, Bd 1. Springer, Berlin Heidelberg, pp 659–662
16. Busse O, Laun A, Agnoli AL (1984) Verschlußhydrozephalus bei Kleinhirninfarkten. Fortschr Neurol Psychiat 52: 164–171
17. Calatayud VP (1985) Zentrale Massenverschiebungen und Zisternenhernien bei akuten traumatischen Hirnläsionen. Dissertation, Gießen 1985
18. Cannon BW (1951) Acute vascular lesions of the brain stem. Complications of supratentorial space-occupying lesions. Arch Neurol Psychiat 66: 687–696
19. Chodkiewicz J, Vedrenne C, Redondo A (1973) Brain stem lesions following head injuries. Clinco-anatomical correlations in 100 cases, ICS 293. Excerpta Medica, Amsterdam, p 69
20. Cioffi FA, Bernini FP et al (1985) Surgical management of acute cerebellar infarction. Acta Neurochir (Wien) 74: 105–112
21. Constant JR, Guerin J et al (1978) CT appearances of the normal tentorial hiatus and expanding lesions of the incisura. J Neuroradiol 5/1: 27–41
22. Cooper PR et al (1979) Traumatically induced brain stem haemorrhage and the computerized tomographic scan: Clinical, pathological and experimental observations. Neurosurg 4: 115–124
23. Crawford JV, Russel DS (1956) Cryptic arteriovenous and venous haematomas of the brain. J Neurol Neurosurg Psychiat 19: 1–11

24. Crompton MR (1971) Brain stem lesions due to closed head injury. The Lancet 1: 669–673

25. Dandy WE (1938) Surgery of the brain. Deutsche Übersetzung: Köbcke H (ed) Hirnchirurgie. J A Barth, Leipzig

26. Del Zoppo GJ, Zeumer H, Marker A (1986) Thrombolytic therapy in stroke: Possibilities and hazards. Stroke 17: 595–607

27. Dhopesh VP et al (1980) Computed tomography in brain stem haemorrhage. J Comput Assist Tomogr 4: 603–607

28. Dinsdale HB (1964) Spontaneous haemorrhage in the posterior fossa. Arch Neurol (Chic) 10: 200–217

29. Doczi T, Thomas DGT (1979) Successful removal of an intrapontine haematoma. J Neurol Neurosurg Psychiat 42: 1058–1061

30. Dongen KJ van, Braakman R, Gelpke GJ (1983) The prognostic value of computerized tomography in comatose head-injured patients. J Neurosurg 59: 951–957

31. Dorndorf W (1983) Schlaganfälle. Thieme, Stuttgart

32. Durward QJ, Barnett HJM, Barr HWK (1982) Presentation and management of mesencephalic haematoma. J Neurosurg 56: 123–127

33. Epstein AW (1951) Primar massive pontine haemorrhage. A clinico-pathological study. J Neuropath Exp Neurol 10: 426–448

34. Feistner H, Busse O, Agnoli AL (1981) Vergleichende klinische und computertomographische Befunde bei Hirnstamminfarkten. Nervenarzt 52: 163–166

35. Gehuchten P van (1937) Le mécanisme de la mort dans certains cas de tumeur cérébrale. Encéphale 2: 113–127

36. George B, Thurel C et al (1981) Frequency of primary brain stem lesions after head injuries. A CT-scan analysis from 186 cases of severe head trauma. Acta Neurochir (Wien) 59: 35–43

37. Graham DI, Adams JH, Doyle D (1978) Ischemic brain damage in fatal non-missile head injuries. J Neurol Sci 39: 213–234

38. Greenacre P (1917) Multiple spontaneous intracerebral haemorrhages. A contribution to the pathology of apoplexy. Johns Hopk Hosp Bull 28: 86–88

39. Groeneveld A, Schaltenbrand G (1927) Ein Fall von Duraendotheliom über der Großhirnhemisphäre mit einer bemerkenswerten Komplikation: Läsion der gekreuzten Pes pedunculi durch Druck auf den Rand des Tentorium. Dtsch Z Nervenheilk 117: 32

40. Haar FL, Sadhu VK et al (1980) Can CT-findings predict intracranial pressure in closed head injury patients? In: Shulman K et al (eds) Intracranial pressure IV. Springer, Berlin Heidelberg New York, pp 48–50

41. Hildebrandt G, Werner M et al (1985) Acute non-communicating hydrocephalus after spontaneous subarachnoid haemorrhage. Acta Neurochir (Wien) 76: 58–61

42. Hinshaw DB, Thompson JR et al (1980) Infarction of the brain stem and cerebellum: a correlation of computed tomography and angiography. Radiology 137:.105–112

43. Ho SU, Kun KS et al (1981) Cerebellar infarction: a clinical and CT study. Surg Neurol 16: 350–352

44. Hounsfield GN (1973) Computerized transverse axial scanning (tomography). Part I: Description of system. Br J Radiol 46: 1016–1022

45. Hounsfield GN (1980) Computed medical imaging (Nobel lecture). J Comput Assist Tomogr 4: 665–674

46. Huhn B, Jakob H (1970) Traumatische Hirnstammläsionen mit vieljähriger Überlebensdauer. Nervenarzt 41: 326–334

47. Jacobs L et al (1976) Autopsy correlations of computerized tomography. Experience with 6000 CT scans. Neurology (Minneap) 26: 1111–1118

48. Jellinger K (1967) Häufigkeit und Pathogenese zentraler Hirnläsionen nach stumpfer Gewalteinwirkung auf den Schädel. Z Nervenheilk (Wien) 25: 223–249

49. Jellinger K, Seitelberger F (1970) Protracted post-traumatic encephalopathy. J Neurol Sci 10: 51–94

50. Jellinger K (1980) Pathology and aetiology of ICH. In: Pia HW et al (eds) Spontaneous intracerebral haematomas. Springer, Berlin Heidelberg New York, pp 13–29

51. Johnson DL, Fritz C et al (1986) Perimesencephalic cistern obliteration: a CT sign of life-threatening shunt failure. J Neurosurg 64: 386–389

52. Kaplan HA (1974) Anatomy and embryology of the arterial system of the forebrain. In: Vinken PJ et al (eds) Handbook of clinical neurology, vol 11. North Holland Publ Comp, Amsterdam New York, pp 1–23

53. Kazner E, Wende S et al (1981) Computertomographie intrakranieller Tumoren. Springer, Berlin Heidelberg New York

54. Kempe LG (1964) Surgical removal of an intramedullary haematoma simulating Wallenberg's syndrome. J Neurol Neurosurg Psychiat 27: 78–80

55. Kernohan J, Woltman HW (1929) Incisura of the crus cerebri due to contralateral brain tumour. Arch Neurol Psychiat 21: 274–287

56. Key A, Retzius G (1875) Studien in der Anatomie des Nervensystems und des Bindegewebes. Samson & Wallin, Stockholm I + II

56a. Kim RC, Fagin K, Choi BH (1985) Prolonged survival after severe traumatic injury limited to the brain stem. Surg Neurol 23: 525–528

57. Kleihues P, Hizawa K (1966) Die Infarkte der A. cerebri posterior: Pathogenese und Beziehungen zur Sehrinde. Arch Psychiat Z Neurol 208: 263–284

58. Knüpling R et al (1979) Chronic and acute transtentorial herniation with tumours of the posterior cranial fossa. Neurochirurgia 22: 9–17

59. Kollikowski H, Ewert T et al (1986) Intraarterielle lokale Fibrinolysetherapie der A. basilaris-Thrombose bei älteren Gefäßpatienten. Akt Neurol 13: 201–206

60. Koos WT, Sunder-Plassmann M, Salah S (1969) Successful removal of a large intrapontine haematoma. Case report. J Neurosurg 31: 690–694

61. Krauland W (1963) Die pathologische Anatomie des Schädel-Hirn-Traumas. Klin Wschr (Wien) 75: 489–492

62. Krauland W (1982) Biomechanik der gedeckten Schädelhirnverletzungen. In: Verletzungen der intrakraniellen Schlagadern. Springer, Berlin Heidelberg New York, pp 20–37

63. Krayenbühl HA, Yaşargil MG (1979) Zerebrale Angiographie für Klinik und Praxis. Thieme, Stuttgart

64. Lanksch W, Kazner E (1978) Schädelhirnverletzungen im Computertomogramm. Springer, Berlin Heidelberg New York

65. Laun A, Agnoli AL (1980) Brain stem haemorrhages. In: Pia HW et al (eds) Spontaneous intracerebral haematomas—Advances in diagnosis and treatment. Springer, Berlin Heidelberg New York, pp 196–201

66. Laun A (1982) Zum Problem der Zisternenverquellung und direkter sowie sekundärer Schädigungen am Hirnstamm (computertomographische Analysen). In: Müller E (ed) Das traumatische Mittelhirnsyndrom. Springer, Berlin Heidelberg New York, pp 60–64

67. Laun A, Agnoli AL et al (1983) Morphological and CT-findings in traumatic brain stem haemorrhages—a contribution of pathophysiology of primary and secondary lesions. In: Villani R (ed) Advances in neurotraumatology, ICS 612. Excerpta Medica, Amsterdam Oxford Princeton

68. Laun A, Busse O et al (1984) Cerebellar infarcts in the area of the supply of the PICA and their surgical treatment. Acta Neurochir (Wien) 71: 295–306

69. Laun A, Agnoli AL, Klug N (1985) Die akute Hypodensität des Hirnstammes im CT — Zeichen des Hirntodes? In: Verhandlungen Dtsch Ges Neurologie, Bd 3. Springer, Berlin Heidelberg New York Tokyo, pp 634–638

70. Lausberg G (1972) Temperaturregulation bei intrakranieller Drucksteigerung. Acta Neurochir (Wien) [Suppl] 19

71. Lindenberg R (1955) Compression of brain arteries as pathogenetic factor for tissue necrosis and their areas of predilection. J Neuropath Exp Neurol 14: 223–243

72. Lindenberg R (1964) Significance of the tentorium in head injuries from blunt forces. Clin Neurosurg 12: 129–142

73. Lindenberg R (1970) Brain stem lesions characteristic of traumatic hyperextension of the head. Arch Path 90: 509–515

74. Lobato RD, Sarabia R et al (1986) Normal computerized tomography scans in severe head injury. J Neurosurg 65: 784–789

75. Lorenz R (1973) Wirkungen intrakranieller raumfordernder Prozesse auf den Verlauf von Blutdruck und Pulsfrequenz. Acta Neurochir (Wien) [Suppl] 20

76. Lui TN, Fairholm DJ (1985) Surgical treatment of spontaneous cerebellar haemorrhage. Surg Neurol 23: 555–558

77. Mahapatra AK, Tandon PN et al (1985) Bilateral decerebration in head injury patients. An analysis of sixty-two cases. Surg Neurol 23: 536–540

78. Mayer ET (1967) Zentrale Hirnstammschäden nach Einwirkung stumpfer Gewalt auf den Schädel. Hirnstammläsionen. Arch Psychiat Neurol 210: 238–262

79. Mayer ET (1969) Zur Pathologie des traumatischen Mittelhirn und apallischen Syndroms. Radiologie 9: 16–22

80. Messina AV, Potts DG et al (1976) Computed tomography: evaluation of the posterior third ventricle. Radiology 119: 581–592

81. Meyer A (1920) Herniation of the brain. Arch Neurol Psychiat 4: 387–400

82. Minauf M, Schacht L (1966) Zentrale Hirnschäden nach Einwirkung stumpfer Gewalt auf den Schädel. II. Mitteilung — Läsionen im Bereich der Stammganglien. Arch Psych Z ges Neurol 208: 162–176

83. Moore HT, Stern K (1938) Vascular lesions in the brain stem and occipital lobe occuring in association with brain tumours. Brain 61: 70–89

84. Murphy A, Teasdale E, Matheson M et al (1983) Relationship between CT indices of brain swelling and intracranial pressure after head injury. In: Ishii S et al (eds) Intracranial pressure V. Springer, Berlin Heidelberg New York, pp 562–565

85. Nagashina C, Watanabe T (1986) Symmetrical thalamic low densities in descending transtentorial herniation. Surg Neurol 25: 29–32

86. Naidich TP et al (1976) Computed tomography in the diagnosis of extra-axial posterior fossa masses. Radiology 120: 333–339

87. Nakagawa Y, Kooyama Y et al (1980) Clinical implications of the CT appearance of the brain stem cistern in cases of acute and severe head injuries. Progr Comp Tomog 2: 235–241

88. Narayan RK et al (1981) Improved confidence of outcome prediction in severe head injury. J Neurosurg 54: 751–762

89. Obrador S, Dierssen G, Odoriz BJ (1970) Surgical evacuation of a pontine-medullary haematoma. Case report. J Neurosurg 33: 82–84

90. Osborn AG (1977) Diagnosis of descending transtentorial herniation by cranial computed tomography. Radiology 123: 93–96

91. Osborn AG, Heaston DK, Wing SD (1978) Diagnosis of ascending transtentorial herniation by cranial computed tomography. Am J Roentgenol 130: 755–760

92. Pak H, Patel SC, Malik GM (1981) Successful evacuation of a pontine haematoma secondary to rupture of a venous angioma. Surg Neurol 15: 164–167

93. Papo I, Pasquini U, Salvolini U (1976) Case reports. Subependymal brain stem haematomas: a report of 2 cases. Neuroradiology 11: 279–282

94. Payne HA, Maravilla KR et al (1978) Recovery from primary pontine haemorrhage. Ann Neurol 4: 557–558

95. Peters G (1966) Morphologische Forschung in der Neurologie und Psychiatrie. Nervenarzt 37: 429–437

96. Peters G (1979) Die Bedeutung der primär und sekundär traumatischen Hirnveränderungen für das klinische Syndrom. Acta Neurochir (Wien) 23: 187–198

97. Peters G, Rothemund E (1977) Neuropathology of the posttraumatic apallic syndrome. In: Dalle Ore G et al (eds) The apallic syndrome. Springer, Berlin Heidelberg New York, pp 78–87

98. Pia HW (1957) Die Schädigungen des Hirnstammes bei den raumfordernden Prozessen des Gehirns. Acta Neurochir (Wien) [Suppl] 4

99. Poppen JL, Kendrick JF, Hicks SP (1952) Brain stem haemorrhages secondary to supratentorial space taking lesions. J Neuropath Exp Neurol 11: 267–279

100. Pullicino P, Kendall BE (1980) Contrast enhancement in ischaemic lesions. Neuroradiology 19: 235–239

101. Riessner D, Zülch KJ (1939) Über die Formveränderungen des Hirns (Massenverschiebungen, Zisternenverquellungen) bei raumbeengenden Prozessen. Dtsch Z Chir 253, 1–61

102. Sager WD, Ladurner G (1979) Klassifikation und Verlauf des Hirninfarktes im Computertomogramm. Fortschr Röntgenstr 131: 470–475

103. Salazar J, Vaquero J et al (1986) Clinical and CT scan assessment of benign versus fatal spontaneous cerebellar haematomas. Acta Neurochir (Wien) 79: 80–86

104. Sano H et al (1982) The range of cerebral microcirculation during increased intracranial pressure. 1st Int Symp: The cerebral veins, Graz

105. Sano K, Ochiai C (1980) Spontaneous brain stem (pontine) haematomas. In: Pia HW et al (eds) Spontaneous intracerebral haematomas. Springer, Berlin Heidelberg New York, pp 156–159

106. Sano K (1983) Spontaneous brain stem haematoma. Neurosurg Rev 6: 71–77

107. Sato S et al (1982) Cerebral microcirculation in experimental sagittal sinus occlusion in dogs. 1st Int Symp: The cerebral veins, Graz

108. Schönmayr R (1985) Die zerebralen Massenverschiebungen — eine Systematik computertomographischer und klinischer Befunde. Habil Schrift, Gießen

109. Schwarz GA, Rosner A (1941) Displacement and herniation of

the hippoicampal gyrus through the incisura tentorii. Arch Psychiat 46: 297 ff

110. Scotti G *et al* (1980) Cerebellar softening. Am Neurol 8: 133–140

111. Scoville WB, Poppen JJ (1949) Intrapeduncular haemorrhage of the brain. Arch Neurol Psychiat 61: 688–694

112. Scriver EB, Olsen TS (1982) Contrast enhancement of cerebral infarcts. Neuroradiology 23: 259–265

113. Seeger W (1968) Atemstörungen bei intrakraniellen Massenverschiebungen. Acta Neurochir (Wien) [Suppl] 17

114. Sellier K, Unterharnscheid F (1963) Mechanik und Pathomorphologie der Hirnschäden nach stumpfer Gewalteinwirkung auf den Schädel. Hefte Unfallheilkunde 76

115. Silverstein A (1972) Primary pontine haemorrhage. In: Vinken PJ, Bruyn GW (eds) Handbook of clinical neurology, vol 12. North Holland Publ Comp, Amsterdam, pp 37–53

116. Silverstein A (1974) Pontine infarction. In: Vinken PJ, Bruyn GW (eds) Handbook of clinical neurology, vol 12. North Holland Publ Comp, Amsterdam, pp 13–36

117. Spatz H, Stroescu GJ (1934) Zur Anatomie und Pathologie der äußeren Liquorräume des Gehirns. Nervenarzt 7: 425–481

118. Stopford JSB (1928) Increased intracranial pressure. Brain 51: 485

119. St John JN, French BN (1986) Traumatic haematomas of the posterior fossa. Surg Neurol 25: 457–466

120. Stovring J (1977) Contralateral temporal horn widening in unilateral supratentorial mass lesions: a diagnostic sign indicating tentorial herniation. J Comput Assist Tomogr 1: 319–323

121. Strich SJ (1956) Diffuse degeneration of the cerebral white matter in severe dementia following head injury. J Neurol Neurosurg Psychiat 19: 163–185

122. Sundaresan N *et al* (1979) Successful surgical treatment of pontine vascular malformation in a 3-year old. Child's Brain 5: 131–136

123. Tabaddor K, Danziger A, Wisoff HS (1982) Estimation of intracranial pressure by CT-scan in closed head trauma. Surg Neurol 18: 212–215

124. Teasdale G, Jennet B (1974) Assessment of coma and impaired consciousness. A practical scale. Lancet 2: 81–83

125. Teilmann K (1953) Haemangiomas of the pons. Arch Neurol Psychiat 69: 208–223

126. Tomaszek DE, Rosner MJ (1985) Cerebellar infarction. Analysis of twenty-one cases. Surg Neurol 24: 223–226

127. Toutant SA, Klauber MR *et al* (1984) Absent or compressed basal cisterns on first CT-scan: ominous predictors of outcome in severe head injury. J Neurosurg 61: 691–694

128. Tsai FY *et al* (1980) CT on brain stem injury. JNR 134: 717–723

129. Tsai FY *et al* (1982) Computed tomography in acute posterior fossa infarcts. AJNR 3: 149–156

130. Unterharnscheid FJ (1972) Die traumatischen Hirnschäden. Mechanogenese, Pathomorphologie und Klinik. Z Rechtsmedizin 71: 153–221

131. Vedrenne C, Chodkiewicz JP (1975) Les lésions du tronc cérébral chez les traumatisés craniens (étude anatomique). Agressologie 16: 1–8

132. Vincent C, David M, Thiébaut F (1936) Le cone de pression temporal dans les tumeurs des hémisphères cérébraux. Rev Neurol (Paris) 65: 536

133. Wackenheim A, Babin E (1978) The midbrain and its displacements in axial computerized tomography. J·Neuroradiol 5/1: 43–56

134. Wilson G, Winkelmann NW (1926) Gross pontile bleeding in traumatic and nontraumatic cerebral lesions. Arch Neurol Psychiat 15: 455

135. Yoshino E, Yamaki T *et al* (1985) Acute brain oedema in fatal head injury: analysis by dynamic CT scanning. J Neurosurg 63: 830–839

136. Zeumer H, Hacke W *et al* (1982) Lokale Fibrinolysetherapie bei Basilaris-Thrombose. Dtsch Med Wschr 107: 728–731

137. Zülch KJ, Mennel HD, Zimmermann V (1974) Intracranial hypertension. In: Vinken PJ, Bruyn GW (eds) Handbook of clinical neurology, vol 6. North Holland Publ Comp. Amsterdam, pp 89–149

Author's address: PD Dr. Albrecht Laun, Department of Neurosurgery, University of Giessen, Klinikstrasse 29, D-6300 Giessen, Federal Republic of Germany.

Acta Neurochirurgica, Suppl. 40, 57–94 (1987)

Electrically Elicited Blink Reflex and Early Acoustic Evoked Potentials in Circumscribed and Diffuse Brain Stem Lesions

Norfrid Klug and **György Csécsei**

Department of Neurosurgery, University of Giessen, Federal Republic of Germany

Contents

1. Introduction and Objectives

Modern radiological investigation techniques permit an increasingly precise imaging of the brain stem and adjacent structures. The imaging of individual structures and their reconstruction in various planes reveals the presence of space-occupying processes, haemorrhages, structural alterations, oedematous swelling and substance defects as density differences or zones of different signal intensity. However, just like conventional X-ray methods, computer tomography and nuclear magnetic resonance tomography does not reveal the extent of a functional disorder in the transverse and longitudinal section of the brain stem. Besides clinical investigation, additional electrophysiological measures are necessary for this purpose. With the aid of these methods, neurological deficits and clinically latent disorders can be collated with pathological alterations in the brain stem. Since the electroencephalogram in brain stem lesions merely shows non-specific alterations, the registration of electrically elicited brain stem reflexes and evoked brain stem potentials have increasing significance, since their generators and connection pathways in the brain stem are known and circumscribed or diffuse functional disorders can be inferred from changes in these potentials. Their registration is simple, can be repeated as often as desired and is not a burden for the patient.

Whereas registration of early acoustic evoked potentials has attained increasing importance in neurological-neurosurgical conditions[14, 15, 20, 30, 41, 52, 53, 59–62, 83, 93, 102–104, 111, 112, 122, 141, 142, 146–149] since the introduction of electronic averagers by Dawson[31, 32], the blink reflex already analysed electromyographically in 1952 by Kugelberg[87] as a very informative diagnostic criterion has hardly penetrated the consciousness of neurosurgeons.

In the present paper, the topodiagnostic value of early acoustic evoked potentials and the blink reflex will be illustrated with reference to a large number of conscious patients and patients with disturbances of consciousness. In addition, its significance in coma monitoring and as a prognostic criterion in patients in acute midbrain syndrome will be shown. Since the development of the acute midbrain syndrome can as a rule not be detected electrophysiologically under clinical conditions, in an experimental part of the study an acute elevation of supratentorial pressure with secondary incarceration of the midbrain was induced in a simulated epidural haematoma in cats. The alterations of the blink reflex and the acoustic evoked potentials occurring during the elevation of intracranial pressure allow an insight into the dynamics of the incipient disturbance of brain stem function which will be discussed in relation to literature data and our own clinical findings.

2. Historical Review

2.1. Blink Reflex (BR)

Overend[117] in 1896 was the first to describe reflex muscular contraction of the orbicularis oculi muscle on tapping the glabella. Kugelberg[87] analysed the reflex electromyographically in 1952. He was able to demonstrate two separate components in mechanical and electrical stimulation of the supraorbital nerve. Later, the reflex pathway was investigated experimentally in cats[161]. The alteration of the late blink reflex response (R_2) in humans was first described in medullary lesions, in damage to the descending nucleus of the trigeminal nerve[75, 87] and in pontine lesions[16, 73, 77, 122]. According to the investigations of other authors, the polysynaptic connections for R_2 reached in the oral direction as far as the level of the nuclei of the facial nerve[34, 113, 155]. However, alterations of R_2 have also been described in mesencephalic vascular damage and mesencephalic m.s. foci[7, 77]. Later, a connection between the lack of R_2 and damage to the mesencephalic reticular formation in decerebrated patients was discussed[126, 135]. Various authors have registered the blink reflex in comatose patients[9, 26, 90, 98]. Some authors regard the blink reflex as an informative prognostic criterion[9, 26, 126].

Our own investigations indicate that the late response of the blink reflex is impaired even in circumscribed mesencephalic lesions[24, 25, 82], so that alterations of R_2 have topodiagnostic and prognostic significance both in chronic and in acute disorders of brain stem function.

2.2. Early Acoustic Evoked Potentials (BAEP)

As early as a few years after the discovery of electroencephalography, potentials could be registered from the human scalp following an acoustic stimulus. By the introduction of averagers[31, 32], the early components of the acoustic evoked potentials generated in the brain stem could be registered far away from the generators of the individual potential waves of the scalp. By means of this "far-field technique", the action potential of the cochlear nerve was registered initially by Sohmer and Feinmesser[138, 139], and later also by other authors[70, 142, 143]. After animal experimental investigations[1, 2, 8, 89], it is assumed that the seven waves generated within 10 msec after a defined acoustic stimulus (these were specified with Roman numbers) arise in the following generators: wave

    I in the cochlea or in the cochlear nerve,
    II in the nucleus of the cochlear nerve,
    III in the medial parts of the superior olive,
    IV in the ventral nucleus of the lateral lemniscus,
    V in the inferior colliculus,
    VI in the medial geniculate body and
    VII in the acoustic radiation.

Registration with deep and surface electrodes in the cat led to the assumption that wave IV is generated in the inferior colliculus and adjacent areas, whereas wave V is likewise assigned to the inferior colliculus or an adjacent area directly rostral to it[69]. On the other hand, nerve division experiments in the cat showed that wave IV was not impaired after ablation of the inferior colliculus[8]. According to recent investigations, it is assumed by some authors that wave II is not generated in the nucleus of the chochlear nerve, but in the cochlear nerve itself[10, 28, 47, 129–131, 140]. These investigations are contradicted by clinical observations which confirm the old theory of the cochlear nucleus area as generator for wave II[68, 81, 82].

The clinical significance of the early acoustic evoked potentials consists in the fact that the corresponding waves of the BAEP and/or the time latency between individual potential peaks (interpeak latency, IPL) may be altered and their time interval may be prolonged in circumscribed lesions which could lead to a functional disorder of the brain stem. On the other hand, it became apparent that the early components of BAEP are stable in changing state of consciousness[4, 70]. They are not affected or not affected to a noteworthy extent by drugs and anesthetics, so that they have an appreciable significance in coma with the clinical signs of brain death and the isoelectric EEG[12, 80–83, 132, 141].

Therapeutic doses of conventional drugs do not affect the BAEP; this is of particular value in differential diagnosis of toxic metabolic causes of global brain stem dysfunction as compared to structural causes. Investigations on drug-intoxicated comatouse patients in whom respiration and brain stem reflexes were absent, showed a normal BAEP[142].

3. Materials and Methods

3.1. Clinical Investigations

Two hundred and thirty-nine patients were investigated, of these 110 with circumscribed space-occupying lesions of the posterior cranial fossa and 129 patients with acute diffuse disturbance of brain stem function. Systemic diseases and degenerative conditions of the central nervous system as well as ischaemic lesions with a primary extracerebral cause were ruled out. Precondition for registration of evoked potentials is an averager enabling summation of the stimulus-related potentials, and elimination of stimulus-independent potentials such as spontaneous EEG, noise and muscle potentials. Furthermore, a stimulus generator, an EEG monitor and a registration unit are needed. For registration of the blink reflex, a two-channel EMG instrument with registration unit is sufficient. The present investigations were carried out with a commercially available DA II R (manufactured by Tönnies) with incorporated artefact suppression and ERA stimulator. An XY plotter served for documentation.

3.1.1. Blink Reflex

Before registration of the blink reflex, the facial nerve is stimulated preauricularly with the bipolar surface electrode (control group n = 27) and the muscle response potential from the orbicularis oculi muscle is registered in order to rule out a disturbance in the peripheral reflex arc, e.g. damage to the facial nerve in fractures of the base of the skull. Stimulation of the supraorbital nerve was carried out with bipolar surface electrodes and the registration from both orbicularis oculi muscles with small surface electrodes or concentric needle electrodes (coaxial needle electrodes, DISA). At least two, but mostly three or more reflex responses were demonstrated on the storage oscilloscope and registered with the XY plotter. The demonstration or the absence of individual reflex responses and (in circumscribed chronic lesions) latency and amplitude values of early response (R_1) and ipsilateral or contralateral late responses (R_2, R_2^C) have been evaluated.

3.1.2. Brain Stem Acoustic Evoked Potentials

BAEP were registered with surface electrodes in accordance with the 10–20 electrode system of the International EEG Federation. Silver chloride electrodes were used. The different electrode was over M 1/2, the "indifferent" electrode was over C_z and the earth electrode was placed in a frontal median location. The registrations were made ipsilaterally to the stimulus. In the control group (n = 35), stimulation was performed with 70 dB SL, i.e. with 70 dB above the subjective auditory threshold (SL = sensory level). In the patient group, stimulation was performed both with 70 dB and with 80 dB HL (HL = hearing level, i.e. with 70–80 dB above the auditory threshold of the control group with normal hearing) in each case. The ear not exposed to the sound stimulus was masked with 60 dB. If an identification of the individual potential components was not possible afterwards, an additional stimulation with 60 dB or 50 dB HL was administered. At least two averaging steps of 2048 single responses in each case were performed in each patient. An alternating click of 200 µsec duration, stimulus sequence 10 Hz served as stimulus. A headphone (Beyer DT 48) was used as sound converter. The electroencephalographic activity was amplified 100,000 times (filter 1 kHz, time constant 10 msec). The latencies of the single potential components are peak latencies which were calculated from the difference between the positive amplitude maximum directed upwards and the beginning of the stimulus. The acoustic signal in the Figures is given 0.2 msec after the beginning of the sequence. In waking patients, sedation with diazepam (Valium[R]) or flunitrazepam (Rohypnol[R]) was regularly administered before the registration.

3.2. Experimental Investigations

The investigations were carried out after intraperitoneal injection of an average of 1.5 ml/kg Narcoren, corresponding to 24 ml/kg pentoarbital sodium in tracheotomized cats fixed in a stereotactic head frame (Fig. 1). The cats were protected from cooling by being placed on a heating cushion and by means of an infrared lamp.

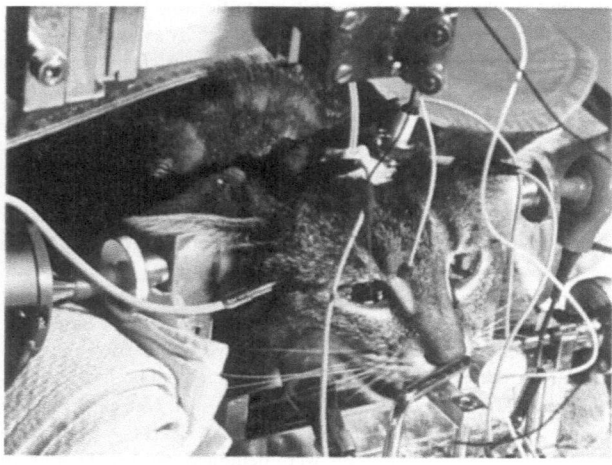

Fig. 1. Experimental paradigm for registration of acoustic evoked potentials and the electrically elicited blink reflex under intracranial pressure elevation

To raise intracranial pressure, a rubber balloon was implanted epidurally in a frontotemporal location on the right and filled with 0.1 ml H$_2$O/min. The intracranial pressure (ICP) was measured at the corresponding position on the opposite side with a pressure transducer from Statham (model P 37 B[R])[170]. Central venous pressure (CVP) and arterial blood pressure (ABP) were determined via catheters in the femoral artery and the femoral vein (Statham SP 50[R]), and the rate of breathing (RR) was determined via a temperature-dependent resistance sensor (thermistor) which was fixed at the exit of the tracheal tube. The heart rate (HR) was measured from the ECG, which was recorded via needle electrodes arranged according to Einthoven. The body temperature was measured rectally (Yellow springs thermoprobe) and subcutaneously with a needle probe (Hartmann and Braun). The EEG was recorded via electrodes screwed into the vault of the cranium fronto-occipitally to it on both sides. The measurements were registered with a 16-channel jet plotter (Mingograph[R], Siemens) with a paper speed of 1.5 mm/sec as well as with a trend plotter (12-channel Kompensograph[R], Siemens).

3.2.1. Blink Reflex

In 10 cats, the blink reflex was elicited by stimulation of the nictitating membrane with two teflon needle electrodes coated up to the tip (electrode distance 5 mm). The reflex responses were registered from the orbicularis oculi muscle on both sides with concentric EMG needle electrodes. After testing the optimal stimulus intensity, no change was made in the conditions of stimulation and registration during the entire investigation.

3.2.2. Acoustic Evoked Potentials

In 12 cats, the BAEP was evoked with a Nicolet stimulator by means of a earphone speaker (Beyer DT 48). The acoustic signal (duration 200 µsec, frequency 10 Hz) was conducted to the auditory duct via a 1.5 cm long conically auricular tunnel via the stereotactic screw drilled out on the inside which was sealed with wax against loss of sound pressure to the outside. Stimulation was carried out with 60 dB to 70 dB above the auditory threshold of the trial manager. For registration of the BAEP, polished EEG needles were inserted subcutaneously on both sides over the mastoid. We used a subcutaneous vertex electrode as the reference electrode. As a rule, the EEG responses to 128 single signals were averaged. A Tönnies DA II R myograph served as averager. In five cats, both the blink reflex and the BAEP were registered.

3.2.3. Investigation of the Blood-Brain Barrier

For investigation of the blood-brain barrier, a 2% Evans blue solution (2 ml/kg body weight) was injected intravenously before the beginning of the first elevation of intracranial pressure. At the end of the experiment, the cats were fixed in 10% formaline solution by intracardial perfusion, and the brain was removed.

4. Results

4.1. Normal Findings

4.1.1. BR

After electrical stimulation of the facial nerve (Fig. 2 a), the direct muscle response potential of the orbicularis oculi muscle was registered ipsilaterally to the stimulus. There may be a bilateral late reflex response (facio-facial reflex). In stimulation of the supraorbital nerve (Fig. 2 b), an ipsilateral early response (R$_1$) and a bilateral late response (R$_2$) appear.

In the control group (n = 27), the latency times of the muscle response potentials (M response) as well as the early and late reflex responses of blink reflex were determined. The following normal values were determined (Table 1).

4.1.2. BAEP

Figure 3 shows the generators of the brain stem acoustic evoked potentials with an original BAEP regis-

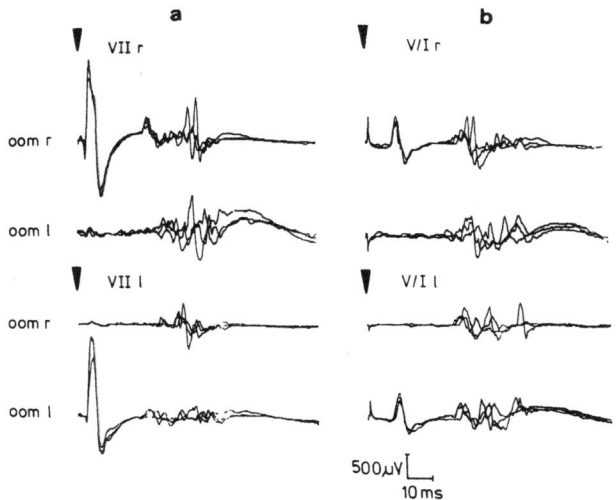

Fig. 2. Normal finding in electrical stimulation of the facial nerve and the supraorbital nerve. a) Stimulation of the facial nerve on the right (VII r.) and the left (VII l.): detection of the stimulus-ipsilateral direct muscle response potential and a bilateral late facio-facial reflex response. b) Stimulation of the supraorbital nerve on the right (V/I r.) and the left (V/I l.): detection of the stimulus-ipsilateral early response (R_1) and a bilateral late response (R_2). *oom r*: musculus orbicularis oculi on the right. *oom l*: musculus orbicularis oculi on the left

tration. Fourfold averaging of 2048 single signals reveals the familiar time constancy of the individual potential components, whereas the amplitudes display a greater variation.

The absolute latency for wave I is 1.89 ± 0.15 msec in our own control group. The absolute latencies of BAEP depend on the stimulus intensity, whereas the interpeak latencies reveal a major intraindividual and

interindividual constancy at different stimulus intensivites[147]. It is therefore usual to determine IPL (as a rule I–III, III–IV, I–V[13, 88, 94, 148, 149] and more rarely also I–II[95, 149]) in the appraisal of BAEP. The interpeak latencies (Table 2) determined in our own control group largely agree with those specified in the literature.

The determination of intraindividual side differences of the interpeak latencies shows only minimal deviations in healthy subjects (Table 3). Side differences which exceed three times the single standard deviations are rated as pathological.

Since there are greater variations in the absolute amplitude of the individual potentials, it is usual to specify amplitude ratios, as a rule with the ratio III/I, V/III and V/I, since these display greater interindividual constancy[142]. Our own normal values are presented in Table 4.

4.2. Circumscribed Processes with Involvement of the Brain Stem

This group comprises 110 patients with space occupations which lead directly or indirectly to brain stem damage. The detection was carried out by means of computer tomography or with nuclear magnetic resonance tomography. The findings were confirmed surgically or (in some cases) by autopsy.

4.2.1. Cerebellar Space Occupations

In 26 patients, there was a cerebellar space occupation which involved one cerebellar hemisphere 21 times and

Table 1. *Latency times of the muscle response potential as well as the early (R_1) and late (R_2) reflex response of the blink reflex.* Control group (n = 27). Data with single standard deviation and mean error

Stimulation / Recording		Latency (ms)	
		R	L
M-Response	R	2,72±0,33 (2,0-3,5)	–
	L	–	2,8±0,35 (2,0-3,6)
R_1	R	10,2±0,54 (10,0-11,8)	–
	L	–	10,1±0,59 (9,5-12,0)
R_2	R	32,2±3,8 (26,0-41,0)	34,2±4,3 (25,0-45,0)
	L	33,1±3,8 (24,0-41,0)	32,5±3,5 (25,0-41,0)

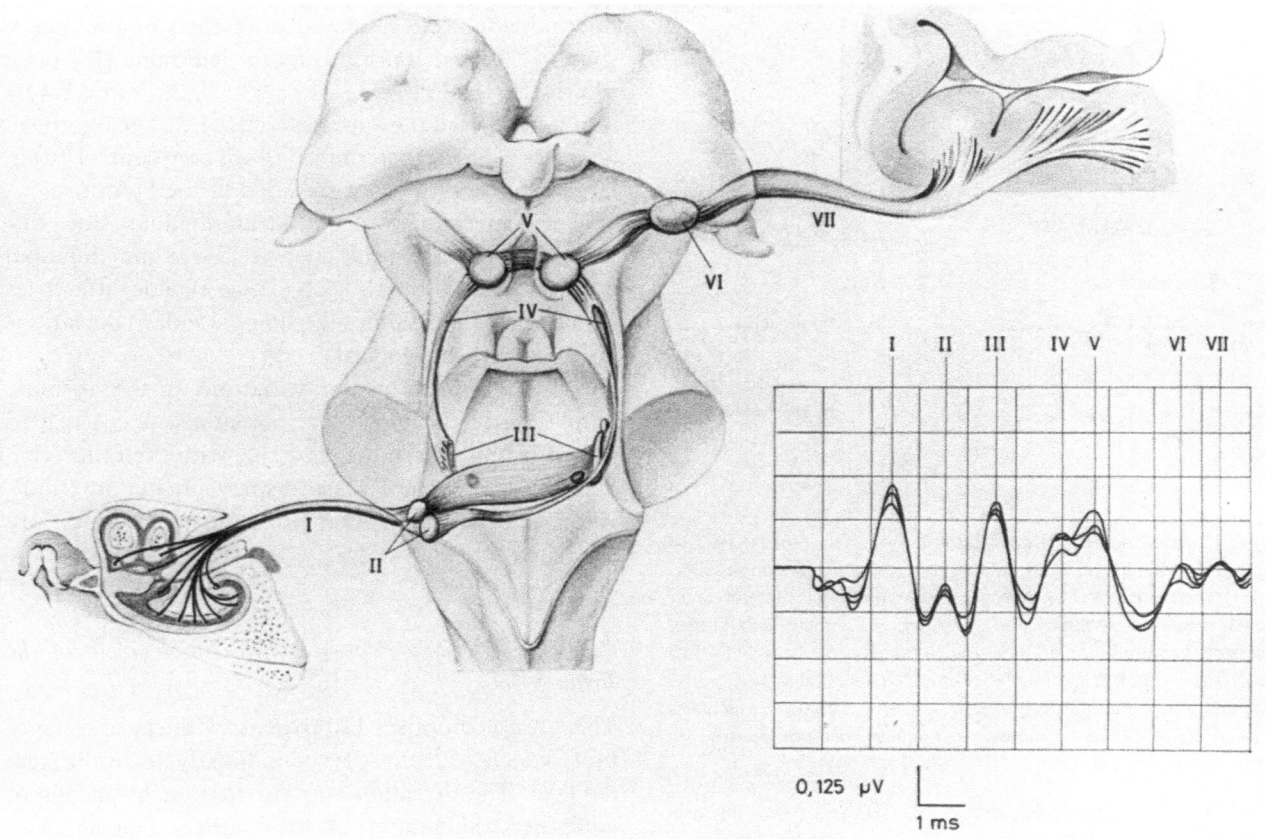

Fig. 3. Schematic representation of the auditory tract with representation of the generators and the individual potential components (I–VII). After a drawing from Nieuwenhuys (1980) and Tandler (1929). BAEP after our own original registration. *I* nerve VIII, *II* nucleus of the cochlear nerve, *III* superior olive, *IV* nucleus of the lateral lemniscus, *V* inferior colliculus, *VI* lateral geniculate body, *VII* acoustic radiation

Table 2. *Mean values of interpeak latencies with single standard deviation (70 dB SL)*

I–II	II–III	V–VI	I–III	III–V	I–V	N
0,98±0,14	1,15±0,11	1,72±0,17	2,13±0,15	1,79±0,12	3,93±0,19	35
69	69	58	70	70	70	

Table 3. *Side differences of the interpeak latencies of the control group (n = 35).* Variations by more than three times the standard deviation (lower column) are regarded as pathological. — = IPL on the right less than the left

	I–II (ms)	II–III (ms)	I–III (ms)	III–V (ms)	I–V (ms)
70 dB SL	-0,003±0,128	-0,027±0,113	-0,029±0,114	0,006±0,127	-0,023±0,128
3 × SD	0,38	0,34	0,34	0,38	0,38

Table 4. *Amplitude ratios of the control group with single standard deviation*

$A_{III/I}$	$A_{V/III}$	$A_{V/I}$	N
$0,91 \pm 0,44$	$1,93 \pm 0,63$	$1,60 \pm 0,51$	35
70	70	70	

Table 5. *IPL in 26 patients with cerebellar space-occupying lesion*

IPL (ms)	Involved Side	Uninvolved Side
I – III $\geq$ 2,6	7	2
III – V $\geq$ 2,1	11	15
I – V $\geq$ 4,5	13	9

the cerebellar vermis five times. In 18 cases, a tumour was involved (seven medulloblastomas, five spongio-blastomas, three metastases, two angioblastomas, one astrocytoma). Six patients had suffered an ischemic cerebellar infarction with subacute clinical symptoms and development of an increasing space-occupying lesion of which the surgical elimination led to an improvement of the clinical state in all cases. One case involved an abscess and one case involved a haematoma of a cerebellar hemisphere.

4.2.1.1. BR

The latency of the direct muscle response after stimulation of the facial nerve was in the normal range on both sides in all patients; in the side comparison, however, the latency was prolonged on the tumour side in five patients. R_1 was negative in two patients, once on the side of the space occupation and once on the opposite side. In six cases, the R_1 latency was prolonged ipsilaterally. In two patients, all late responses were absent after stimulation of the supraorbital nerve on both sides. We have seen a loss of R_2 and R_2^c only after stimulation of the "healthy" side in one case. The latency of R_2 was prolonged on both sides in five patients, only on the intact side in three further patients and on the pathological side in one patient.

In five patients, there was a prolonged latency for R_2^c on both sides, after stimulation of the healthy side in only three cases and after the stimulation of the "diseased" side in two cases. An "asynchronous" R_2 and an inconstant behaviour of latency and amplitude was frequent.

4.2.1.2. BAEP

We saw a loss of all BAEP waves on the pathological side in two patients with medulloblastoma. We were able to observe a disintegration of one or several potential peaks and/or the absence of more than two potential components in four further patients. A pro-

longation (more than 3 SD) of the interpeak latencies of different brain stem segments was striking (Table 5). The Table shows that the functional disorder most frequently involved the pontomesencephalic segment on the tumour-free side. A displacement of the brain stem to the opposite side owing to the tumour may explain this. The pathological blink reflex findings contralateral to cerebellar space occupations can be explained in a similar way (Fig. 4).

4.2.2. Cerebellopontine Angle Tumours

Fifty-seven patients with cerebellopontine angle tumours (comprising 37 acoustic neurinomas and 14 meningiomas, six cases with metastases, chondromyxomas and cholesteatomas) were investigated.

4.2.2.1. BR

The blink reflex was normal or the R_1 latency on the tumour side was at the upper range of normal in six patients. We did not see an absence of both reflex response on the neurinoma side. The most frequent and striking result was a pathologically prolonged R_1 latency on the tumour side in 31 cases. The amplitude of R_1 was frequently, but not regularly reduced. The R_1 latency was normal or prolonged in small neurinomas without brain stem compression. In large extrameatal neurinomas with compression of the pons, the R_1 latency on the tumour side might amount to double or more than double the contralateral latency (Fig. 5), often only with a slight reduction in the amplitude compared to the "healthy" side. We saw a loss of the ipsilateral R_1 four times in large neurinomas.

Alterations of R_2 also occurred in large neurinomas which had led to compression of descending trigeminal nuclei owing to tonsillar herniation. On the other hand, there were pathological R_2 alterations in blockade of the aquaeduct with obstructive hydrocephalus. In large neurinomas and stimulation on the tumour side, the

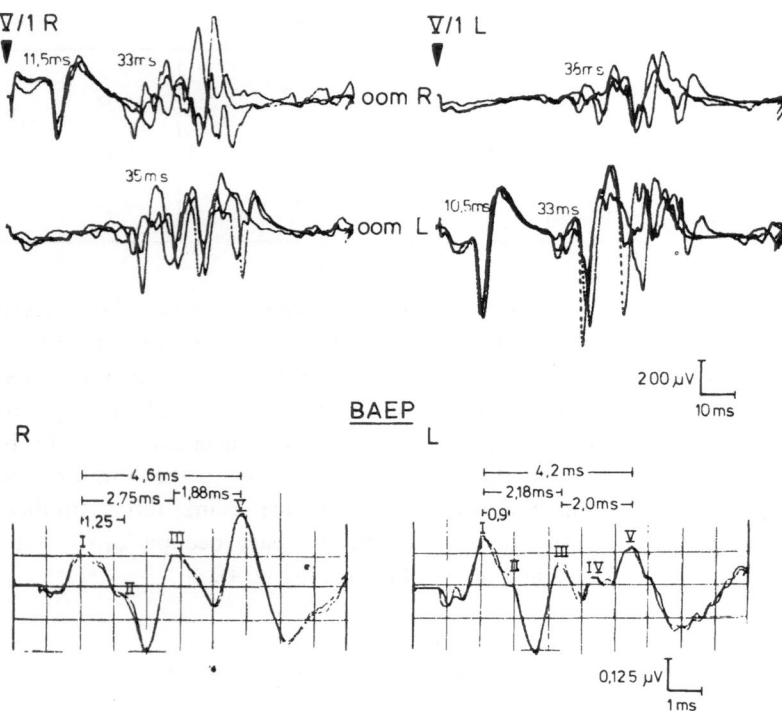

Fig. 4. D.I., 43 years old, female: right cerebellar metastasis. Preoperative BR (top) and BAEP (bottom) registrations

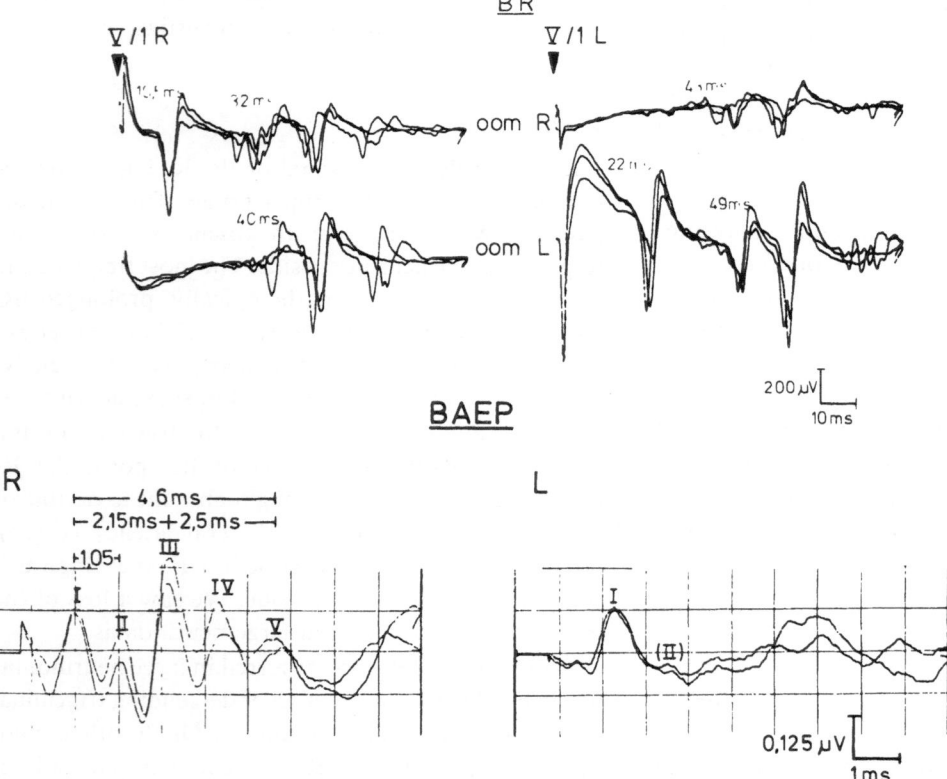

Fig. 5. H.E., 44 years old, female: large acusticus neurinoma on the left side. BR and BAEP registrations. Clinical loss of hearing on the left side. No facial nerve paresis

ipsilateral R_2 latency could be longer than the contralateral R_2 latency. Absence of R_2 did not occur.

4.2.2.2. BAEP

In the acousticus neurinomas, all tumour-ipsilateral and the majority of tumour-contralateral BAEP registrations were pathological, whereas the blink reflex was normal or the tumour-ipsilateral R_1 latency was at the upper limit of normal in six patients. However, we have also seen a difference in the latency in more than one msec between sides in these patients. In 24 patients, all BAEP waves on the tumour side were absent, and the BAEP was pathological on the neurinoma side in the remaining 13 patients. The most striking point in all patients was a side-different BAEP, apart from one female patient with bilateral acousticus neurinoma who had loss of the potential chain on both sides. The patients with BAEP loss on the tumour side also displayed contralateral alterations with emphasis on the waves IV and V. The alterations depended on the tumour size: in small intrameatal tumours with slight extrameatal growth, only the BAEP on the tumour side was pathological, whereas in large extrameatal neurinomas with compression of the pons, there were also delays of the contralateral interpeak latencies (especially IPL III–IV) besides the alterations on the same side. Large neurinomas with compression and displacement of the brain stem to the opposite side had frequently led to loss of all waves on the tumour side and to severe deformation of the IV/V complex up to contralateral loss of wave V (Fig. 5). In contrast to the acousticus neurinomas, the BAEP was absent on the tumour side only once in the 20 remaining cerebellopontine angle tumours. In this case of an extensive cholesteatoma which had led to destruction of the inner ear, the contralateral BAEP was normal.

4.2.3. Space-Occupying Processes of the Brain Stem

A circumscribed space occupation in the brain stem was present in 27 patients comprising 22 tumours, two circumscribed haemorrhages and three brain stem angiomas. On the basis of the clinical symptoms (cranial nerve involvement) and the computer tomographic, intraoperative or autopsy findings, the lesion was mesodiencephalic in three cases, purely mesencephalic in four cases, pontomesencephalic in five cases, pontine in six cases, pontobulbar in four cases and bulbar in one case. The mesencephalon, pons and medulla oblongata were damaged in one patient, and in three patients a tumour of the fourth ventricle had also

infiltrated the brain stem. In two cases, the process was strictly localized in the midline with corresponding symmetrical symptoms, whereas a certain side dominance could be demonstrated clinically and radiologically in the other patients. In these patients, we have distinguished between a "pathological" and "healthy" side for didactic reasons.

4.2.3.1. Blink Reflex

4.2.3.1.1. M Response

The facial M response on the damaged side was absent in two patients; in these patients, R_1 and R_2 were also absent on the pathological side, whereas R_2^c could be elicited. The latency of the M response was prolonged on the pathological side compared to contralateral in only one patient. The duration of the M response on the tumour side was prolonged in six patients (Fig. 6), and there was a reduction of amplitude in five patients.

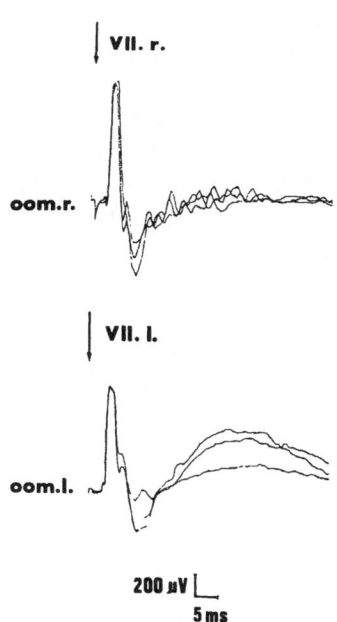

VII. r.

oom.r.

VII. l.

oom.l.

200 µV

5 ms

Fig. 6. S. B., 15 years old, female: pontobulbar glioma, more pronounced on the right. Prolonged M response on the tumour side

4.2.3.1.2. R_1 Response

We saw a loss of R_1 on the lesion side in 10 patients; these also had clinical signs of pontine damage (including peripheral facial paresis). The R_1 latency was prolonged on the pathological side in five patients. An amplitude reduction on the pathological side was found in 16 patients, and a bilateral reduction of amplitude in one patient.

4.2.3.1.3. R_2 Response

We saw a bilateral loss of R_2 in six cases, including two patients with a pontobulbar lesion and four patients with a pontomesencephalic or mesencephalic lesion. Four patients showed unilateral loss of R_2.

In two patients with pontomedullary lesion, the stimulus-ipsilateral R_2 was absent, whereas the stimulus-contralateral R_2 was absent in two mesencephalic tumours. A bilaterally delayed latency of R_2 was present in seven cases, and a unilateral latency delay of R_2 was present in 10 patients. The unilateral or bilateral prolongation of R_2 was striking in pons tumours: this exceeded 500 msec in one patient (Fig. 7). We have seen an amplitude reduction of R_2 in 14 patients. R_2^c was lost on both sides in six patients, and it was absent five times on the pathological side alone (in stimulation on the healthy side). A bilateral latency prolongation was found eight times, and a unilateral latency prolongation six times. The duration of R_2 was increased bilaterally in one patient and unilaterally in four patients. Unilateral amplitude reduction was present in seven cases. Independently of the localization, we have considered the following alterations as typical:

1. In mesodiencephalic and mesencephalic processes, the M response and R_2 were not appreciably altered. As a rule, R_2 was pathological (always the stimulus-contralateral response in lesion on a specific side).

2. In pontomesencephalic or oral pontine lesions, R_1 on the tumour side was mostly pathological or absent. R_2 was pathological in tumour-contralateral or bilateral stimulation. The latency and duration of R_2 were prolonged—loss of R_2 was rare.

3. In lower pontine and pontobulbar tumours, R_1 usually had a pathological delay of latency, reduced amplitude and was frequently abolished on the lesion side. R_2 was likewise pathological, mostly in stimulation on the tumour side, but also in bilateral stimulation (Fig. 8).

4. In tumours of the 4th ventricle (three patients), we have seen a bilateral amplitude reduction of R_1 responses and a bilateral loss of R_2 in one patient. In the other two patients, the latency and duration of R_2 were likewise prolonged on both sides.

4.2.3.2. BAEP

The BAEP showed pathological changes in all 27 patients. There were the following specific findings:

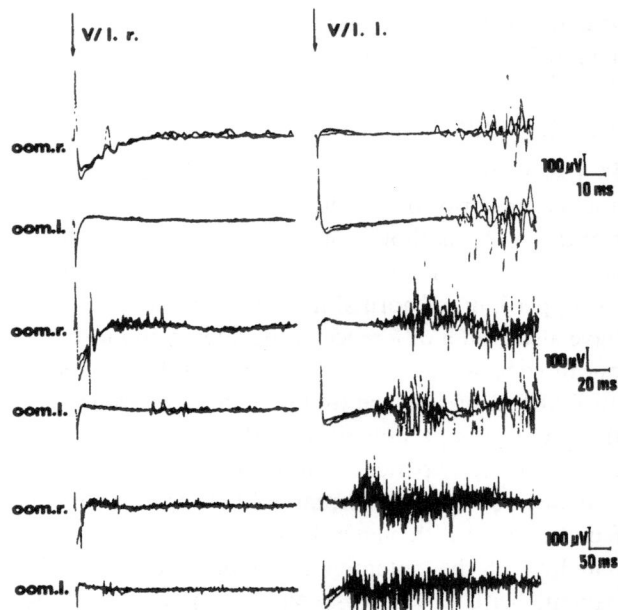

Fig. 7. F.O., 60 years old, male: pontomesencephalic tumour on the left. BR recordings with different speed of registration. Duration of R_2 in stimulation on the left side is extremely prolonged

1. A bilateral absence of all potentials in two patients: the one case was a circumscribed pontomesencephalic haemorrhage concentrated on the right side and with clinically preserved hearing on both sides. The absence of R_1 on the right side and the bilateral absence of R_2 indicated the severe disorder of brain stem function in the conscious patient. The second patient was a two year old child with a bulbar tumour, clinically with severe hypacusia and unable to walk. The R_1 responses could be elicited in this patient, whereas R_2 was absent on both sides.

2. Unilateral absence of BAEP on one side and pathological waves contralaterally in three pons tumours. In two of these, R_1 and R_2 was also absent on the pathological side, and the amplitude of R_1 was reduced and the duration of R_2 increased in one patient.

3. A unilateral lack of potentials generated rostrally from wave I in three pontobulbar and one mesencephalopontobulbar tumour.

4. A bilateral desynchronization of all potentials generated rostrally from wave II in a pontobulbar tegmental tumour (Fig. 8). The blink reflex of this patient showed a latency-prolonged, amplitude-reduced R_1 on both sides, with pathologically altered R_2 responses. In a further 10 patients, all potentials could be registered, although there were appreciable fluctuations in the latencies, IPL and amplitude conditions, and the morphological structure of the individ-

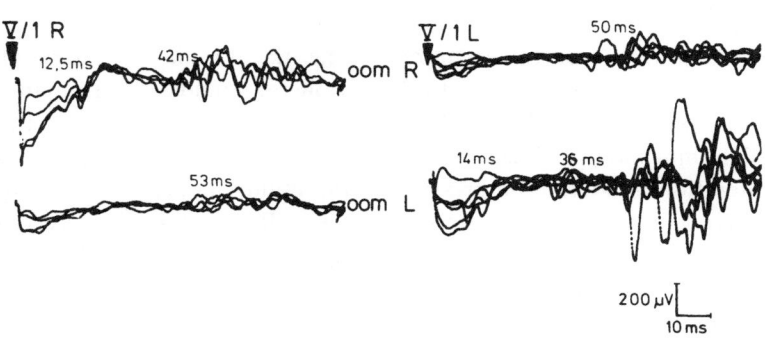

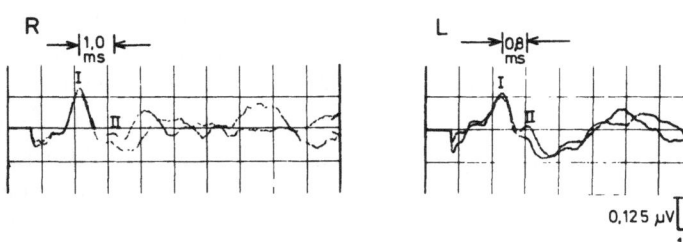

Fig. 8. E.K.H., 37 years old, male: pons glioma. Clinically and radiologically, both sides were affected. BR and BAEP registrations show pathological alterations on both sides

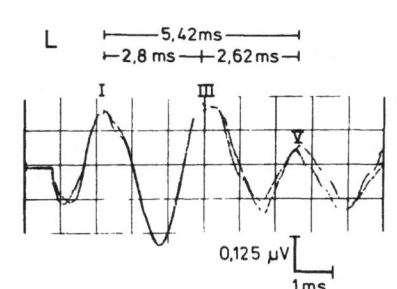

Fig. 9. R.M., 3 years old, male: medulloblastoma of the 4th ventricle with appreciable compression of the brain stem. Pathological BAEP on both sides

ual waves was greatly altered in some cases. Three of these patients had a mesencephalic, one a pontomesencephalic, three a pontine, one a pontobulbar and two an intraventricular space occupation (Fig. 9).

5. A bilateral absence of wave V in three side-dominant pontomesencephalic tumours. The pronounced latency and amplitude instability was striking. In each of these patients, R_1 was absent on the pathological side. The latency and duration of the R_2 responses were prolonged.

6. Desynchronized potentials with prolonged interpeak latencies in three mesodiencephalic and one mesencephalic tumour. An amplitude reduction of wave II was striking in these rostrally situated tumours.

4.2.4. Discussion

In the literature, only occasional studies were con-

cerned with blink reflex findings in space-occupying intracerebral or extracerebral lesions in the region of the posterior fossa or of the cerebellopontine angle[40, 58, 76, 91, 120, 124]. We have not found specific alterations in the patients with cerebellar space occupations. The pathologically altered R_1 and/or R_2 responses (occasionally also on the tumour-contralateral side) might be explained by direct pressure on the trigeminal roots or descending trigeminal nuclei as well as a compression of the reticular formation of the lower brain stem. One might also discuss an indirect action of pressure following obstructive hydrocephalus. The registration of the blink reflex in the diagnosis of acousticus neurinoma or in cerebellopontine angle tumours has particular significance. Even very small intrameatal acousticus neurinomas already lead to damage of the facial nerve. The pathological latency prolongation of the R_1 responses with normal facial M responses was striking in our patients. The reason for this is that in the case of stimulation of the supraorbital nerve, the facial nerve is stimulated transsynaptically, i.e. the stimulation takes place proximally to the peripheral damage. Consequently, there is an alteration of R_1 even in very slight morphological damage. Another cause of the R_1 alteration is lateral compression of the middle pons, where the reflex center is localized. The R_2 latency alterations on the pathological side are located at the damage to the facial nerve (efferent damage), whereas bilateral R_2 alteration is due to compression of the medulla oblongata, i.e. damage to the descending trigeminal nucleus area (afferent damage). There is a detailed discussion of our own blink reflex findings in Section 4.2.5. with regard to the significance of the pontomesencephalic structures in the development of R_2.

The literature references on BAEP changes in tumours of the posterior fossa, especially in acousticus neurinomas, are extensive[11, 17–19, 38, 39, 84, 93, 94, 102–104, 109, 118, 127, 133, 134, 137, 139, 142–144]. The increased latency of wave V[18, 19, 133, 160] is regarded as a criterion for differential diagnosis. Particular significance is attached to the interaural latency of wave V (ILV)[18, 133]. Other authors refer to the increased interpeak latency I–V[19, 39] and suggest that the IPL I–V be established by means of wave I determined in electrocochleography in presence of a negative wave I on the tumour side[39]. A lack of all waves of BAEP on the neurinoma side has been reported by various authors[39, 93, 118, 133, 134]. In our own material, the absence of all waves on the tumour side was a very frequent finding. It is to be noted that the BAEP after sedation was recorded once more when the appraisal was not unaequivocal, and that only artefact-free curves were evaluated. In examination of the corresponding literature, sedation does not appear to be accorded this importance. The BAEP artefacts which are frequently to be observed precisely in extracerebral space-occupying lesions of the posterior cranial fossa might be explained by muscle contracture or the altered signal-to-noise ratio. The loss of wave I in acousticus neurinoma might be due to pressure of the tumour on the cochlear nerve with oedematous reaction, axonal destruction and demyelination[137]. Our findings on contralateral BAEP alterations are largely consistent with communications of other authors[93–95, 109, 118, 133, 137, 168]. This might be explained by damage to the fibers of the auditory tract decussating in the trapezoid body in small tumours, whereas a displacement of the brain stem with contralateral compression

can result in alterations in wave V or the IV/V complex and a prolongation of the IPL III–V in extensive tumours. According to our own findings, these contralateral alterations have special significance, especially from the point of view of topodiagnostics. On the other hand, the major diagnostic criterion in blink reflex is the latency delay R_1 on the tumour side. In our experience, the frequently pronounced latency delay in well-preserved R_1 indicates disturbed function in the region of the caudal pons, but on the other hand it renders probable the extracerebral localization of the tumour. In connection with the bilaterally asymmetrical BAEP, which without exception showed greater pathological alterations on the side of the tumour in our patients, the diagnosis of an extracerebral space-occupying process in the cerebellopontine angle can be largely substantiated.

In contrast to the bilateral asymmetrical BR and BAEP in extracerebral processes, intracerebral brain stem tumours show bilateral symmetrical alterations whereas intracerebral midline lesions of the posterior cranial fossa growing especially on the side can give rise to difficulties in differential diagnosis compared to cerebellopontine angle processes[83].

4.2.5. The Significance of Pontomesencephalic Structures in the Development of the Late Component of BR (R_2)

In preserved R_1 (i.e. in no noteworthy disturbance in the function of the lower pons), we have observed a bilateral loss of R_2 both in medullary and in pontomesencephalic tumours (4.2.3.1.3.). We have likewise only found a unilateral loss or latency prolongation of R_2 both in bulbar lesions concentrated on a specific side and in asymmetrical pontomesencephalic lesions. For this purpose, we have broken down our patients with bilaterally normal R_1 and pathologically altered R_2 into two groups, and evaluated the R_2 latencies statistically. The first group comprises patients with circumscribed lesions caudal to the facial nucleus area (Table 6) and the other group consists of patients with pontomesencephalic or mesencephalic lesions which were unaequivocally rostral to the facial nuclei (Table 7).

There was a distinct difference between these two groups: in the medullary processes, R_2 was pathologically altered on the damaged side, whereas there was a pathological absence of R_2 on the contralateral side to the damage in the pontomesencephalic lesions, i.e. in stimulation on the "nonpathological" side. Findings which we have obtained in a patient with a circum-

Table 6. *Variations of the R_2 latencies in 23 patients with circumscribed medullary damage*

N	Involved Side		Uninvolved Side	
	R_2	$R_2{}^c$	R_2	$R_2{}^c$
15	$46,0\pm5,9$***	$44,5\pm5,9$**	$40,9\pm4,8$	$43,3\pm5,7$
1	-	-	(+)	(+)
1	-	(+)	(+)	-
1	-	-	(+)	-
1	-	-	-	(+)
4	-	-	-	-
Total 23	15	16	18	17

***: p < 0,001
** : p < 0,01   (+): pathologic response
* : p < 0,02   - : no response

Table 7. *Alterations of the R_2 latencies in 17 patients with circumscribed pontomesencephalic damage*

N	Involved Side		Uninvolved Side	
	R_2	$R_2{}^c$	R_2	$R_2{}^c$
7	$45,3\pm14,2$	$42,6\pm7,0$	$52,6\pm16,1$**	$51,6\pm4,2$*
3	(+)**	(+)**	-	-
1	(+)	(+)	(+)	-
1	(+)	-	(+)	(+)
5	-	-	-	-
Total 17	12	11	9	8

**: p < 0,01   + : normal response
* : p < 0,02   (+): pathologic response
                      - : no response

scribed mesencephalic haemorrhage indicate the involvement of mesencephalic structures in the genesis of R_2.

Case report (B., K., 45 years old, male)

The patient suffered a closed craniocerebral trauma with primary loss of consciousness on October 29th 1986. After emergency treatment (intubation, controlled ventilation), he received medical intensive care. In the computer tomogram, subarachnoid blood was found in the region of the perimesencephalic cisterns with a circumscribed mesencephalic haematoma on the left side. The patient could be extubated on the next day. Nine days after the trauma, he was somnolent, and there was amnesia for the accident event. An anisocoria with moderately mydriatic pupil on the right and a miotic pupil on the left, weak light reaction on both sides and slight ptosis on the left side were found neurologically. The corneal reflex on the right side was weakened. In addition, there was a slight hemiparesis and hemihypesthesia including the area innervated by the trigeminal nerve and a slight hemiataxia on the right as well as a dysarthria.

Nuclear magnetic resonance tomography on the 13th day after the trauma showed an isolated signal-intensive zone situated left mesencephalically (Fig. 10). Further alterations could be demonstrated neither in the computer tomogram, nor in nuclear magnetic resonance tomography. The patient could be discharged on the 29th day after trauma with a slight dysdiadochokinesis, hemiataxia and hypesthesia on the right side. The registration of the blink reflex on the ninth day after the trauma showed a maintained R_1 on both sides and a loss of all late responses in stimulation on the right side. In stimulation on the left side, R_2 was shown to be almost normal ipsilateral to the stimulus, whereas it was reduced in amplitude and had a delayed latency contralateral to the stimulus ($R_2{}^c$) (Fig. 11). The result was almost unchanged on the 22nd day. On the 27th day, R_2 and $R_2{}^c$ showed a certain recovery in stimulation on the right side.

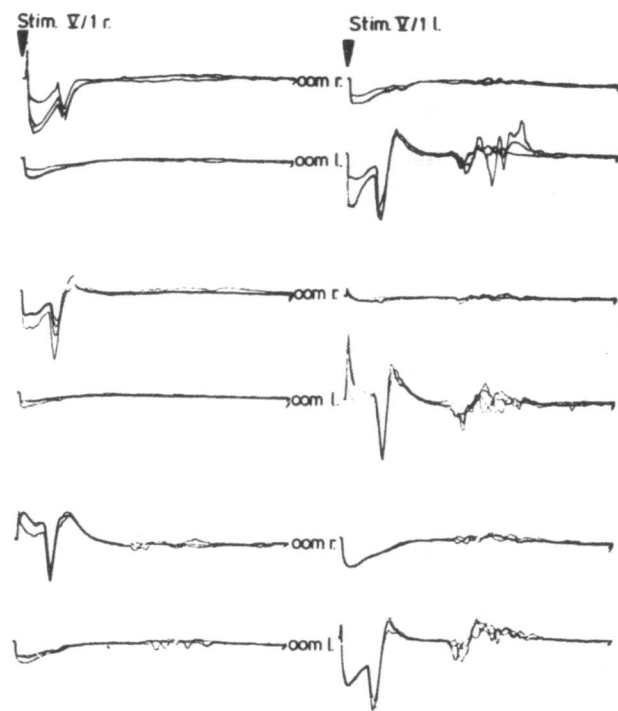

Fig. 11. BR registrations in a circumscribed left mesencephalic traumatic haematoma on the 9th day (top), 22nd day (middle) and 27th day (bottom) after the trauma. Same calibration as in Fig. 12

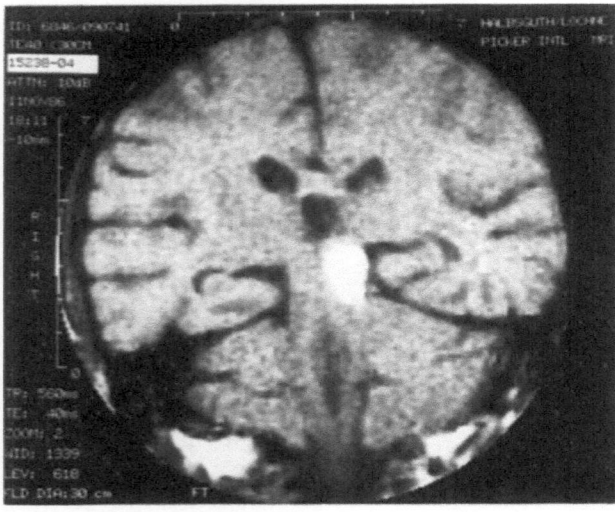

Fig. 10. Nuclear magnetic resonance tomography on the 13th day after the trauma shows a circumscribed mesencephalic increase of signal intensity on the left. See text for further details

4.2.5.1. Discussion

It is generally recognized that a bilateral normal R_1 of the blink reflex indicates maintained function of the caudal pons. Alterations of R_2 in normal R_1 can hence only arise in disorders of rostral (upper pons, mesencephalon) or caudal (medulla oblongata) structures. The involvement of the medulla and in particular of the descending trigeminal nucleus in the genesis of R_2 is proved. This is where the synapsing of the afferent neurons takes place. However, the further pathway of the reflex tract up to the facial nuclei has not yet been clarified. In the present case, only slight damage to the motor and sensory pathways is found clinically on the ninth day after the craniocerebral trauma. At this time, computer tomography and nuclear magnetic resonance tomography showed a left mesencephalic lesion in the form of a circumscribed haemorrhage. The contralateral R_2 could only be demonstrated on the 27th day

after the trauma. Its late recovery indicates a reversible functional disorder in the region of the mesencephalon on the left side, possibly due to oedema. Both the radiological and the clinical findings indicate that the mesencephalon on the right side is morphologically and functionally intact.

In a similar way, Brainin et al.[8] found a complete irreversible loss of the contralateral R_2 in an isolated spontaneous tectotegmental mesencephalic haemorrhage. In this case, as well as in the case we have reported here, the pyramidal tract was clinically uninvolved. The blink reflex findings present in our case correspond to the pathological alterations of type B described by Ongerboer de Visser[115] and by Hopf[67], which are ascribed in the literature to bulbar lesions. In this case, the pathological R_2 is always ipsilateral to the damage. On the other hand, in the strictly lateralized mesencephalic lesions described by ourselves and other authors, R_2 is always pathological contralateral to the lesion.

After unilateral labelling with horseradish peroxidase of the intermedial part of the facial nucleus in which the origin of the motoneurons of the orbicularis oculi muscles is located[21], direct links to the descending nucleus of the trigeminal nerve and to the ventromedial bulbar reticular formation on both sides were found. A bilateral concentration was found which was more pronounced contralaterally in the pontine and mesencephalic tegmentum as well as in the colliculus and in the superior olive, in the oculomotor nucleus and in the red nucleus[156, 157]. The collicular and olivary links are probably responsible for the optic and acoustic blink reflexes. In contrast to humans, blinking in animals is organized by complex mechanisms: it is a combination of retraction (retractor muscle of the eyeball, abducens nerve), upward deviation (Bell's phenomenon, oculomotor nerve) and lid closure (facial nerve). This explains the extensive connections of the facial nucleï in the entire region of the reticular formation of the lower brain stem[63]. On the basis of horseradish peroxidase labelling and autoradiographic investigations, the postulation of concrete reflex pathways is not justified. Ogasawara[110] asserts that the late response of the blink reflex in the cat is trisynaptic. If this were also the case in humans, then the intermediary neuron would have to be very long on the basis of latency time of R_2; one might thus explain the involvement of pontomesencephalic structures in the genesis of the late response. On the basis of our own results[25] and observations of other authors, we consider that the integrity both of stimulus-ipsilateral medullary and stimulus-contralateral pontomesencephalic structures is indispensable for the genesis of a normal late response of the blink reflex in humans. Further investigations are required for precise determination of the reflex pathway.

4.3. Diffuse Acute Processes with Involvement of the Brain Stem

4.3.1. Blink Reflex

Investigations were carried out in 71 patients in acute midbrain syndrome (AMS), in bulbar syndrome and the stage of brain death. Forty-seven patients were in AMS. The cause was a craniocerebral trauma (11 primary and 14 secondary brain stem lesions in 25 cases), a spontaneous intracerebral massive haemorrhage in eight cases, an incisural herniation in supertentorial tumours in seven cases and a subarachnoid haemorrhage in four further cases (three aneurysmatic haemorrhages and one brain stem haemorrhage in AV angioma). In two cases, there was a basilar artery thrombosis and in one case an acute occlusion of the internal carotid artery. Table 8 shows a global view of the patients.

4.3.1.1. Blink Reflex Findings in Acute Midbrain Syndrome

All 47 patients in acute midbrain syndrome had a pathological blink reflex. The facial stimulation carried out at the beginning of the registration revealed a positive M response on both sides in each case. Depending on the degree of alteration of the blink reflex, the patients were divided into three groups:

a) R_1 and R_2 positive, latency times pathologically prolonged (n = 22),

b) detection only of R_1 (n = 14),

c) loss of all reflex responses (n = 11).

Spontaneous breathing was maintained in all patients. The muscle tonus was enhanced (group a) and b)) or decreased (c)). The corneal reflex was absent in all cases with negative R_1 and R_2. It could be elicited in five patients with only positive R_1 and in 18 patients with detection of R_1 and R_2. A positive light reaction was present 18 times in a), three times in b) and only once in c). Of 22 patients with the most favourable BR finding, 13 survived, of 14 patients with only positive R_1, six patients survived and of 11 patients with loss of all reflex responses only one patient survived (Table 9).

Table 8. *Synoptic overview of the clinical state and course of 71 patients with brain stem damage*

SYNDROME	N	IMPROVED	APALLIC	✝
Acute midbrain syndrome	47	15	5	27
Bulbar syndrome	16	–	–	16
Brain death	8	–	–	8
Total	71	15	5	51

Table 9. *Blink reflex results, neurological symptoms and clinical course of 38 patients in acute midbrain syndrome*

BR-FINDING	N	RESPIRATION	TONE	CORNEAL-REFLEX	PUPILLARY LIGHT REACTION	ALIVE	⊥
a) $R_1 + R_2$ pos.	22	(+)	(+)↑	18+ 4-	18+ 4-	13	9
b) only R_1 pos.	14	(+)	(+)↑	5+ 9-	3+ 11-	6	8
c) $R_1 + R_2$ neg.	11	(+)	(+)↓	11-	1+ 10-	1	10
Total	47					20	27

(+) in some cases pathologically detectable; ↑ raised; ↓ reduced

Regular checks of the blink reflex reflect the clinical course. This is to illustrated with a case description:

H.M., 22 years old, male

Multiple trauma with closed craniocerebral trauma. Coma, insufficient spontaneous breathing, gaze turned to the right. Upper decerebration syndrome. In the EEG, left centroparietotemporal focus. In the computer tomogram, detection of free perimesencephalic cisterns. In the first registration (18.5.1982), a prolonged, amplitude-reduced and latency-delayed R_2 and a doubtful contralateral R_2 (R_2^c) was found only in stimulation of the left side. Three days later (21.5.1982), there was a nonspecific pain defense for the first time, and a further ten days later there was an increased clearing of consciousness with opening of the eyes on calling (1.6.1982) (Fig. 12). At this time, the late responses of the blink reflex can be elicited for the first time, whereas an inconstant amplitude-reduced and latency-delayed R_1 can be demonstrated. A further six days later, the first verbal expressions are registered. These are accompanied by a further normalization in the blink reflex. After rehabilitation (14.12.1982), the neurological finding was normal. Only a marked psychopathological syndrome of organic origin was found.

Of the 11 patients in whom all reflex responses were absent only one ten year old child survived with severe defect healing. The absence of both R_2 responses in the first examination indicates a primary mesencephalic lesion, whereas the loss of both R_1 responses documents an extensive pontine damage.

4.3.1.2. BR Findings in Apallics, in Bulbar Syndrome and Brain Death

Six patients were investigated in the stage of the apallic syndrome, 16 further patients in the bulbar syndrome and eight patients in brain death (Table 10). All apallics had an early reflex response, whereas the late response was positive in only one. A latency instability and distortion of individual reflex response was striking in this patient (Fig. 13). The bulbar patients (preserved, spontaneously insufficient breathing, decline of muscular tonus, mydriatic fixed pupils, abolished corneal

Table 10. *Blink reflex findings in apallics, in bulbar syndrome and brain death*

SYNDROME	NUMBER OF PATIENTS	M	R_1	R_2	IMPROVED	⊥
Apallic	6	6	6	5	4	1
Bulbar	16	16	0	0	0	16
Brain death	8	8	0	0	0	8

M = muscle response in facial nerve stimulation

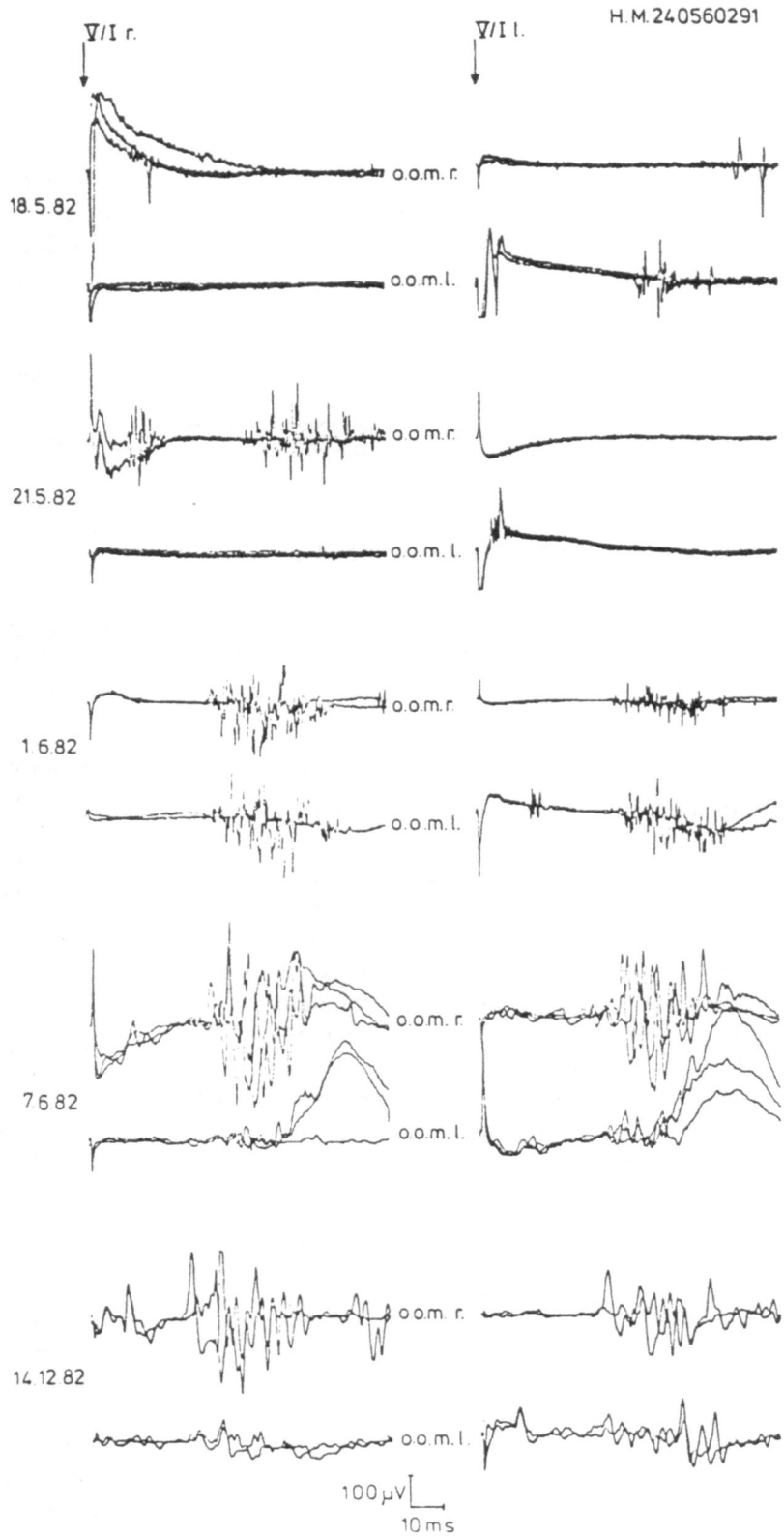

Fig. 12. H.M., 22 years old, male: progress control of the blink reflex in closed craniocerebral trauma with primary brain stem damage. Regression of the late responses corresponding to the clearing of consciousness. Normalization of the initially negative reflex responses on both sides. Note: the first three registrations were made with needle electrodes

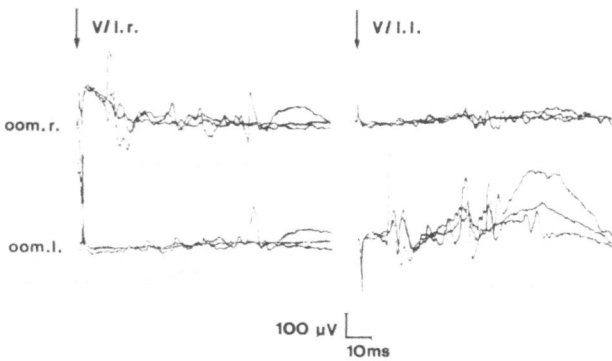

Fig. 13. A.B., 21 years old, male: apallic syndrome. Condition after craniocerebral trauma and relief of a bilateral subdural hygroma. Condition after decerebration syndrome. BR two months after the trauma

reflex, absent deep tendon reflexes) had positive M responses only, whereas R_1 and R_2 were absent. In brain death, all reflex responses were likewise lacking: as a rule, the muscle response potential was positive, but might also be absent.

4.3.1.3. The Blink Reflex as Prognostic Criterion

In the patients with primary and secondary brain stem damage, some typical features could be detected. The most important difference was that the alterations were mostly symmetrical in secondary brain stem lesions, whereas there were mainly asymmetrical alterations in primary lesions. In some cases, there were surprising findings, *e.g.* absent R_1 responses in maintained R_2. In other patients, it was not possible to delimit R_1 and R_2 from each other (Fig. 14). On the basis of the blink reflex findings described, a graduation from the 0–V was applied; the M response could be elicited on both sides in every case.

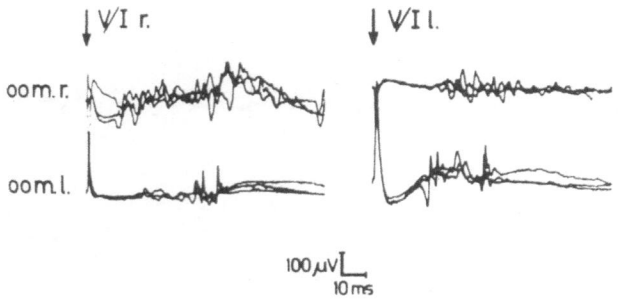

Fig. 14. B.R., 22 years old, male: condition after craniocerebral trauma with primary brain stem contusion and decerebration syndrome. After stimulation of the supraorbital nerves, R_1 and R_2 cannot be separated from each other

Grade 0: R_1 and R_2 are negative on both sides. This finding is the manifestation of a very severe disorder of mesencephalic and pontomedullary function.

Grade I: R_1 can only be elicited on one side, and there is possibly detection of a contralateral R_2 in addition (one or two responses positive). This finding indicates a severe primary brain stem lesion, but also arises from indirect damage, *e.g.* in pronounced elevation of supratentorial pressure.

Grade II: Bilateral absence of R_2 (only two reflex responses positive). This findings is typical for secondary lesions.

Grade III: Pathologically altered reflex responses, unilateral absence of R_1 and/or R_2 (three to five reflex responses can be elicited). This finding indicates a minor pontine and/or mesencephalic disorder.

Grade IV: R_1 and R_2 are present with a delayed latency and reduced amplitude on both sides (six responses positive). This finding only occurred in exceptional cases in AMS.

Grade V: R_1 and R_2 unremarkable on both sides. This result corresponds to a normal finding.

The patients were broken down once more in accordance with the described graduation (Table 11). The good correlation between the degree of severity of BR and clinical course which can be seen from the Table allows a prognostic appraisal which may possibly be improved further by inclusion of additional parameters (latency, amplitude and duration of the single reflex responses).

4.3.1.4. Discussion

Lyon *et al.*[90] was the first to point out the significance of the blink reflex in comatose patients. In patients with coma without indication of herniation in the tentorial cleft, they found a normal R_1. On the other hand, the first reflex response was negative or was delayed in severe transtentorial herniation with haemorrhage in the pons and in coma owing to primary brain stem haemorrhage, brain stem infarction or anoxia. The correlation of the depth of coma with the blink reflex results revealed that the prognosis is very much more favorable in detection of R_2 than in its absence[9]. The return of R_2 correlated strictly with the pathoanatomical mesodiencephalic coma stage which constitutes a level above which the prognosis is favorable in a high percentage of cases. Various authors consider that loss of R_2 or its return after passage of the coma has prognostic significance[135]. In a disturbance of higher cerebral functions, there is a loss of habituation of R_2;

Table 11. *Blink reflex findings on the basis of the graduation described at the time of the first investigation and clinical course of 71 patients with brain stem damage*

GRADING	N	PRIMARY	SECONDARY	IMPROVED	⸱∓⸱
IV–V	–	–	–	–	–
III	18	18	–	12	6
II	8	–	8	4	4
I	10	7	3	3	7
0	35	16	19	1	34
TOTAL	71	41	30	20	51

the return of habituation thus constitutes a good sign for the restoration of cerebral functions, since the integrity of cortical structures which have inhibitory effects on brain stem structures is necessary for this. Some authors point out the possibility of using the blink reflex as an additional criterion in the diagnosis of brain death[98]. According to our own investigations, caution appears to be called for in this regard, since all reflex responses were absent 11 times among 47 patients in acute midbrain syndrome. Simultaneous use of BR and BAEP raises the diagnostic and prognostic certainty in acute midbrain syndrome. In absence of the blink reflex, the acoustic evoked brain stem potentials may serve to demonstrate maintained brain stem function. In failure of BAEP (e.g. in primary labyrinthine deafness, in injury to the cochlea or the acoustic meatus by fractures of the base of the skull), the blink reflex allows an appraisal of the extent and localization of the brain stem lesion.

4.3.2. BAEP

In 110 patients with acute brain stem lesions, 343 BAEP registrations were made. 79 patients were investigated in acute midbrain syndrome, four in apallic syndrome, 35 in bulbar syndrome and 55 in brain death. Special attention was paid to patients in acute midbrain syndrome.

4.3.2.1. BAEP in Acute Midbrain Syndrome

4.3.2.1.1. Interpeak Latencies

The conduction time I–II and V–VI was normal in the first 15 days of acute midbrain syndrome. On the other hand, IPL II–III and III–V and the central (I–V) conduction time were significantly prolonged in the first 48 hours. These findings indicate a functional disorder of the auditory tract in the region of the medulla oblongata, pons and mesencephalon, whereas the segment rostral to the inferior colliculus is normal. From the third day, there is a regression of the pontomesencephalic latency whereas the medullopontine segments remain prolonged up to the 15th day (Table 12).

A greater functional impairment of the caudal brain stem in the first days of the acute midbrain syndrome is manifested in these findings. The results indicate that the regression of the acute midbrain syndrome is initially accompanied by a normalization of disturbed function of rostral brain stem segments.

4.3.2.1.2. Amplitude Ratios

The amplitude ratio III/I and V/I is significantly reduced in the first 10 days (p less than 0.01). This finding persists up to the 15th day, with only an alteration in the significance level for III/I (Table 13). The amplitude ratio A V/III shows a slight rise in the first four days. However, a comparison of the absolute amplitudes of III and V with the control group shows that A III and A V are substantially reduced with values of $0.13 \pm 0.084\,\mu V$ (normal $0.19 \pm 0.07\,\mu V$) and $0.22 \pm 0.12\,\mu V$ (normal $0.39 \pm 0.095\,\mu V$) respectively. On the fifth and sixth day, the amplitude A III falls further to $0.9 \pm 0.055\,\mu V$, whereas A V rose to $0.26 \pm 0.11\,\mu V$. From the divergent development of the amplitude A III and A V at the cost of the pontine

Table 12. *IPL findings in the first 15 days of the acute midbrain syndrome as compared to the control group*

	I-II	II-III	V-VI	I-III	III-V	I-V	DAY	n
CONTROL GROUP	0,98 ± 0,14 69	1,15 ± 0,11 69	1,72 ± 0,17 58	2,13 ± 0,15 70	1,79 ± 0,12 70	3,93 ± 0,19 70	-	35
PATIENTS	1,05 ± 0,47 96	1,45 ± 0,28[++] 84	1,63 ± 0,34 30	2,43 ± 0,31[++] 100	2,00 ± 0,30[++] 88	4,42 ± 0,45[++] 90	1. /2.	52
	0,99 ± 0,21 65	1,53 ± 0,27[++] 57	1,73 ± 0,23 20	2,52 ± 0,36[++] 81	1,92 ± 0,34 66	4,41 ± 0,46[++] 72	3. /4.	34
	1,03 ± 0,29 32	1,49 ± 0,20[++] 25	1,81 ± 0,20 14	2,58 ± 0,28[++] 43	1,84 ± 0,25 36	4,46 ± 0,35[++] 42	5. /6.	24
	0,94 ± 0,15 24	1,36 ± 0,20[++] 21	1,61 ± 0,09 5	2,34 ± 0,25[++] 28	1,91 ± 0,26[+] 27	4,29 ± 0,40[++] 29	7.-10.	13
	0,99 ± 0,18 6	1,47 ± 0,24[++] 6	1,93 ± 0,28 3	2,49 ± 0,32[++] 14	1,81 ± 0,26 14	4,30 ± 0,29[++] 14	11.-15.	8

+ p less than 0.05; + + p less than 0.01

Table 13. *Amplitude ratios in the first 15 days of acute midbrain syndrome*

	$A_{III/I}$	$A_{V/III}$	$A_{V/I}$	Day	N
Control Group	0,91 ± 0,44 70	1,93 ± 0,63 70	1,60 ± 0,51 70	-	35
PATIENTS	0,67 ± 0,36* 97	2,04 ± 1,23 81	1,14 ± 0,63** 83	1./2.	52
	0,51 ± 0,31** 77	2,42 ± 1,52 64	1,05 ± 0,57** 70	3./4.	34
	0,38 ± 0,26** 40	4,33 ± 4,47** 34	1,17 ± 0,60** 35	5./6.	24
	0,36 ± 0,24** 28	4,22 ± 2,73** 27	1,11 ± 0,47** 27	7.-10.	13
	0,59 ± 0,34* 14	2,09 ± 0,76 13	1,10 ± 0,50** 13	11.-15.	8

* p less than 0.05; ** p less than 0.01

generators in the pons and the mesencephalon. In accordance with the IPL results, the amplitude behaviour shows the beginning of a normalization in the mesencephalic part of the auditory tract.

potential component, a significant rise of V/III results (p less than 0.01).

The alterations of the amplitudes and the amplitude ratio indicate a functional impairment of the BAEP

4.3.2.1.3. "Morphological" Structure of Individual Potential Components and Latency Instability

In accordance with the alterations of interpeak latencies and amplitude ratios described, there are alterations of the individual potential peaks in terms of desynchronization, deformation and flattening in 96% of the curves evaluated (Fig. 15). In contrast to chronic

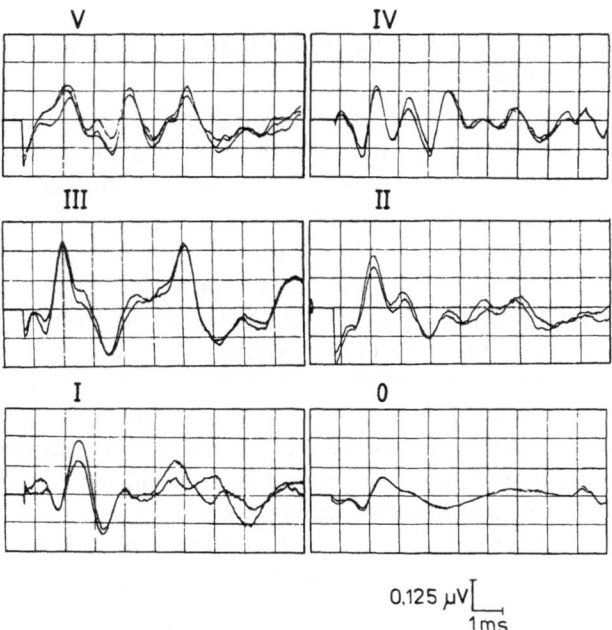

Fig. 15. BAEP-graduation in the acute midbrain syndrome

processes, there is latency instability of individual potential components in 34% of the patients, i.e. there were latency shifts of individual potential peaks with synchronous wave I in the registration period (3–9 min). This "latency instability" is rated as pathological when it persists in registrations with different stimulus intensities (50–80 dB HL). The latency and IPL instability is most pronounced within the first days of the acute midbrain syndrome (Fig. 16). The BAEP findings presented in Figure 16 with secondary rise of the latencies and interpeak latencies with simultaneous fall in amplitude and subsequent gradual normalization correlated with the clinical course. The latency instability already observed before the second deterioration corresponds to that which is typically to be seen in the first days of acute midbrain syndrome.

4.3.2.1.4. BAEP, Lesion Level and Neurological Brain Stem Symptoms

The findings discussed in the past sections indicate a disturbed function of the auditory tract at all levels of the brain stem in the acute midbrain syndrome. The latency delay of individual components described by other authors in neurological diseases[36, 79, 94, 95, 142, 143] is hence only of limited significance for precise localization of the damage in acute midbrain syndrome. However, the results presented document that the central point of the functional disorder can be determined.

Comparing and contrasting neurological and BAEP findings in acute midbrain syndrome leads to the following observations: in bilateral zero-BAEP or only positive wave I, spontaneous breathing was always abolished (exception: one patient with bilateral anacusia after acute basilar artery occlusion). All clinical signs of brain death were present. In additional de-

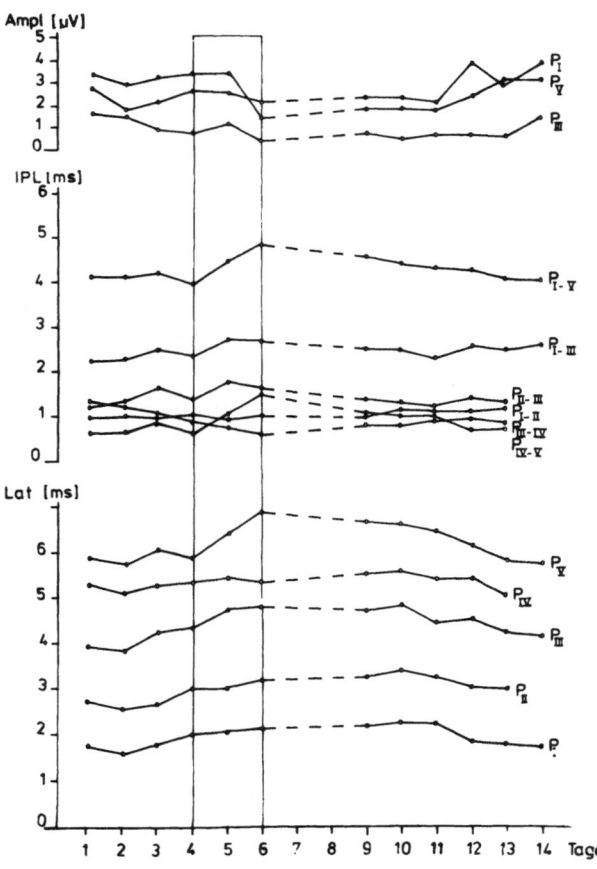

Fig. 16. B.R., 22 years old, male: curve of amplitudes, latencies and interpeak latencies in closed craniocerebral trauma with primary brain stem lesion. Secondary deterioration with computer tomographic detection of a tonsillar herniation on the fourth day. Coma, miotic pupils, light reaction negative, decerebration, GCS 4. Stabilization from the seventh day. Appreciable fluctuation of latencies, interpeak latencies and amplitudes. The column marks the secondary deterioration

tection of wave II, spontaneous breathing was insufficient in 13 out of 16 patients and was absent in three cases; muscle tonus was flaccid or reduced. There was absence of automatic horizontal eye movements, the corneal reflex and the light reaction. In positive wave I–III, spontaneous breathing was always present, but muscle tonus was reduced or increased. Automatic horizontal eye movements occurred in four patients. The corneal reflex was attenuated or absent, and the same applied to the light reaction. The patients with positive waves I–IV all had maintained spontaneous breathing, and frequently a tachypnea. Muscle tonus was regularly raised. Automatic horizontal bulbar movements occurred in seven cases. Corneal reflex and light reaction might be absent (Table 14).

4.3.2.1.5. BAEP Findings Differing on the Right and Left Side

Differences in the interpeak latencies between the right and the left side (difference more than 3 SD) were found in 9 patients. The neurological brain stem symptoms were consistent with the IPL prolongation more pronounced on one side in six cases. In delay of waves III–V, the corneal reflex was abolished ipsilaterally in two cases. Four patients had mydriasis on the side of the IPL prolongation which involved the segments I–III, III–IV and I–V in one case each; in one case, I–III and I–V were prolonged. In the seventh patient, a computer-tomographically demonstrated pons haematoma which is more pronounced on the left side was consistent with ipsilaterally prolonged waves II–III. In one case, no side difference was found neurologically

and by computer tomography in different interpeak latencies. One patient with prolonged IPL III–V which were more pronounced on one side had an attenuated corneal reflex contralaterally. In three patients, a side comparison could not be performed because of peripheral damage (petrosal bone fracture, haematotympanon) with loss of BAEP.

4.3.2.1.6. BAEP Findings in Bulbar Syndrome and Brain Death

The connection between neurological findings and BAEP alterations was already described (Table 14). Here, it is to be discussed specifically for the bulbar syndrome and brain death.

Of 21 patients with loss of the potential peaks rostral wave III, desynchronization of wave III and maintained wave I/II, 12 showed a transition from decerebration into the bulbar syndrome with decrease of muscular tonus, dilated (mostly non-reactive) pupils and increasing loss of the corneal reflex and horizontal eye movements in maintained breathing. Among 16 patients with positive waves I and II and loss of all subsequent potential components, muscle tonus was flaccid 13 times and reduced three times. Corneal reflex, pupillary reaction and horizontal eye movements were absent. Spontaneous respiration was present 13 times and abolished three times (Table 15).

In 55 patients with clinical signs of brain death including isoelectric EEG, the BAEP was registered. In 20 cases, there was a zero-BAEP on both sides. Of the remaining 35 patients, 25 displayed a positive wave I on

Table 14. *BAEP findings and local brain stem symptoms in acute midbrain and bulbar syndrome as well as in brain death*

BAEP-FINDING	N	RESPIRATION	TONE	⟷	CORNEAL-REFLEX	PUPILLARY LIGHT REACTION
No BAEP	20	–	–	–	–	–
Only P I pos.	35	–	–	–	–	–
P I–II pos.	16	3– 13(+)	13– 3↓	–	–	–
P I–III pos.	21	21(+)	12↓ 9↑	4+	13(+) 8–	6(+) 15–
P I–V pos.	38	38(+)	38↑	7+	23(+)	21(+)

↔ = automatic horizontal eye movements; — = negative; + = positive; (+) = positive, in some cases pathological; ↑ = raised; ↓ = attenuated; N = number of patients.

Table 15. *BAEP findings and neurological symptoms in bulbar syndrome*

BAEP	N	RESPIRATION	TONE	⟶	CORNEAL-REFLEX	PUPILLARY LIGHT REACTION
P I-II(III) pos.	12	12(+)	12↓	1+ 3-	4(+) 8-	2(+) 10-
P I-II pos.	16	13(+) 3-	3↑ 13-	-	-	-

Table 16. *BAEP findings in brain death*

	N	%	⊦±⊦
P I bilat. pos.	25	45,4	25
P I unilat. pos.	10	18,2	10
P I pos. total	35	63,6	35
P I bilat. neg.	20	36,4	20
Total	55	100,0	55

both sides, 10 further patients displayed a wave I on one side and a zero-BAEP on the opposite side (Table 16). In all patients in whom potential components beyond wave I had been shown in the preliminary registrations, a loss of wave II was shown with the manifest respiratory arrest. The final result with demonstration of wave I alone or "zero-BAEP" was always aequivalent to clinical brain death. The cerebral panangiography carried out on 10 patients in the context of transplantation surgery documented the intracranial circulatory arrest.

4.3.2.1.7. The Prognostic Significance of BAEP

Repeated BAEP registrations allow timely detection of a secondary deterioration of primary brain stem lesions and secondary involvement of the brain stem in primary cerebral processes. Because of the fluctuating BAEP findings in the first days of acute midbrain syndrome, on the one hand the diagnostic relevance of the single registration is restricted; on the other hand, the interpeak latencies and the amplitude ratios are still within the physiological variation even in major fluctuations, so that the individual finding still has major importance. With consideration of the alterations of interpeak latencies, amplitude ratios, desynchronization and change in form of individual potential peaks which have been described, the first registration of each patient is assigned the degree of severity of a BAEP graduation and correlated with the clinical course.

The following graduation has been used:

Grade 0: only wave I or "zero" BAEP still detected as a common finding in brain death.

Grade I: desynchronisation III, absence of IV and V, IPL I–II is usually prolonged. Grade I constitutes the most severe pontomesencephalic damage at the transition to the bulbar syndrom.

Grade II: prolonged IPL I–III and/or III–V, desynchronisation from wave III. It comprises an extensive pontine and mesencephalic functional disorder including superior olivary level.

Grade III: loss or pronounced alteration of wave III or IV/V, with possibly prolonged IPL I–III and/or III–V. This finding reveals to circumscribed pontine or mesencephalic lesions.

Grade IV: prolonged IPL I–III and/or III–V, possible incipient desynchronisation IV/V. The finding entails a functional pontomesencephalic disorder especially effecting the oral brain stem.

Grade V: corresponds to the normal finding (Fig. 15).

If the BAEP grade and clinical course are correlated, there is a more unfavorable prognosis with decreasing degree of severity. No patient in acute midbrain syndrome had a normal finding. Grade I patients (transition decerebration/bulbar syndrome and incomplete bulbar syndrome with maintained breathing, mydriatic fixed pupils, reduction in tonus and reflexes) all died. Grade 0 patients already fulfilled the clinical criteria of brain death during the registration (Table 17).

Table 17. *BAEP findings and outcome in 71 patients with acute brain stem lesions*

Grade	N	survived	⊹
V	–	–	–
IV	10	8	2
III	15	7	8
II	22	5	17
I	16	–	16
0	8	–	8
total	71	20	51

4.3.2.1.8. Discussion

The registration of brain stem acoustic evoked potentials in patients with acute midbrain lesion has proved to be a valuable diagnostic and prognostic method in our own investigation, and is consistent with communications in the literature on BAEP alterations in the acute phase after craniocerebral trauma[108, 132, 136, 162, 164]. Follow-up registrations allow the effect of therapeutic measures to be checked[80, 132]. The pronounced latency instability of individual potential components especially in the first days of the acute midbrain

syndrome is evidently a manifestation of the brain stem which is characterized on the one hand by spontaneous processes (electrolyte shifts, swelling, circulation, hyperventilation) and on the other hand by therapeutic measures (controlled ventilation, antiedematous treatment, improvement of microcirculation, compensation of the water and electrolyte balance). The alteration of amplitudes, amplitude ratios and interpeak latencies mainly demonstrate a functional disorder both of the oral and of the caudal brain stem during the acute midbrain syndrome. During the clinical stabilization or regression of AMS, the normalization begins first of all in the pontomesencephalic segments and later in the medullopontine segments.

Various suggestions have been made for graduation of BAEP in post-traumatic coma[52, 53, 108, 136, 164]. The graduation in grade I to grade IV suggested by Greenberg *et al.*[52] appears questionable to the extent that I corresponds to a normal finding and II with an absence of the wave VI and VII has no significance, since waves VI and VII may already be absent under normal conditions[13, 14, 93, 94]. A subdivision into two groups, the first with delayed latency V or loss of V and a second with loss of waves I–V[162] is likely to be too imprecise for clinical use. A subdivision of comatouse patients into five types[164] entails an amplitude reduction of wave I and a loss of the remaining potential components for type I. In type II, waves I and II are normal, waves III and V are absent. Type III is characterized by an intact wave I and II in alterations of waves III–V. Type IV displays a normal I–II and altered IV and V. A normal BAEP corresponds to type V. The authors themselves consider that type IV is questionable, because they also classified patients in this type whose BAEP merely displayed a widening or change in form of the potential components IV/V.

Our own graduation considers alterations of amplitudes/amplitude ratios and interpeak latencies as well as alterations in the form of individual potential components. The consideration of all factors already enables a prognostic appraisal after a single registration. The findings obtained at brain death confirm that BAEP has special significance in the evaluation of brain death[14, 80, 132, 141].

4.4. Experimental Findings

In the evaluation of the experimental findings, it is to be noted that the axis of the brain stem of the cat is almost straight, and the area of the tentorium is steeper than in humans. During a controlled rise in intracranial pressure, there is hence an earlier axial displacement of the brain stem and thus bulbar or combined lesions in animals than in humans. This explains why frontal tumours act on the medulla oblongata in animals with a slight rise in pressure, whereas temporal tumours only act in the presence of a larger increase of ICP.

4.4.1. Normal Findings

4.4.1.1. Blink Reflex

Under the registration conditions described under "Methods", a blink reflex can be regularly demonstrated in the cat with stimulus-ipsilateral early response (R_1) and bilateral late response (R_2). The stimulus-ipsilateral late response takes place earlier than the contralateral R_2. The latency times of all reflex responses are shorter than in humans (Fig. 17).

Before the beginning of intracranial pressure elevation, the optimal stimulus intensity to trigger the blink reflex is determined. The dependence of the R_2 and R_1 latencies on the stimulus intensity is shown in Figure 18.

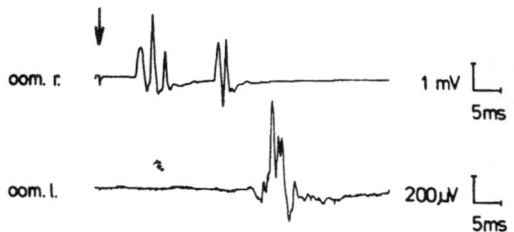

Fig. 17. Normal blink reflex in the lightly anesthesized cat

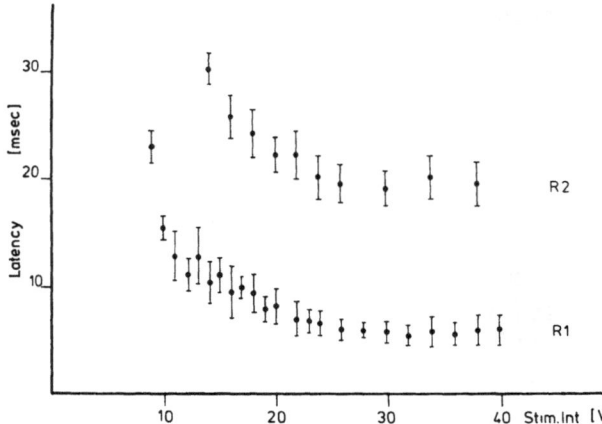

Fig. 18. Latency dependance of the early (R_1) and late (R_2) blink reflex response on stimulus intensity

4.4.1.2. Acoustic Evoked Potentials

Registrations with surface and deep electrodes[70, 89] as well as nerve division experiments[8] in the cat led to demonstration of five potential peaks within the first 10 msec after an acoustic stimulus. In contrast to humans (wave V), wave IV regularly shows the largest amplitudes in the cat. The absolute latencies and the interpeak latencies are shorter, the amplitudes I, III and IV are relatively high and largely constant intra-individually under anesthesia (Fig. 19).

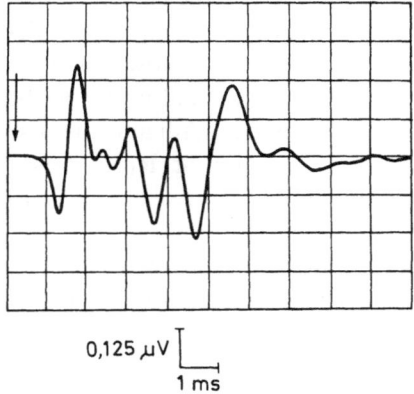

Fig. 19. BAEP normal findings in the anesthesized cat (alternating monaural clicks, ipsilateral registration with needle electrodes)

4.4.2. Results During the Elevation of Intracranial Pressure

4.4.2.1. Pathophysiological Findings

During the filling of the epidurally implanted balloon, there is an initial rise of intracranial pressure corresponding to the extent of simulated epidural haematoma. There were no noteworthy alterations of blood pressure and central venous pressure, and there was a slight decrease of breathing rate only in a few cats. With a further increase of the space occupation, the ICP showed a fluctuating plateau. In this phase, there was a more or less severe tachypnea in almost all cats, followed by periodic breathing, fall of heart rate and a rise of systolic arterial and central venous blood pressure in all animals. At this time, mydriasis occurred on both sides, beginning as a rule on the side of the space occupation. With further filling of the balloon, there was respiratory arrest, fall of systemic arterial blood pressure and central venous pressure as well as the heart rate in parallel with a fall of ICP after reaching ICP values of 80–100 mm Hg. The pupils were maximally dilated on both sides and without light reactions. The animals died within a short time.

4.4.2.2. Blink Reflex

During the rise of intracranial pressure, there was a shortening of the latency of both reflex responses up to ICP values of 30–40 mm Hg. In addition, six out of ten cats showed an increase in amplitude of R_1 and R_2. With a further increase of pressure, the latencies of all reflex responses increased and the amplitudes decreased. The first reflex response to be lost was R_2 at ICP values of about 50 mm Hg; it could no longer be demonstrated in the further course. With a further rise of pressure, latency and duration of R_1 increased appreciably, whereas the amplitude fell and R_1 was finally suppressed completely (Fig. 20 a).

After the relief of pressure carried out at this time, R_1 recovered rapidly, but latency and amplitude remained

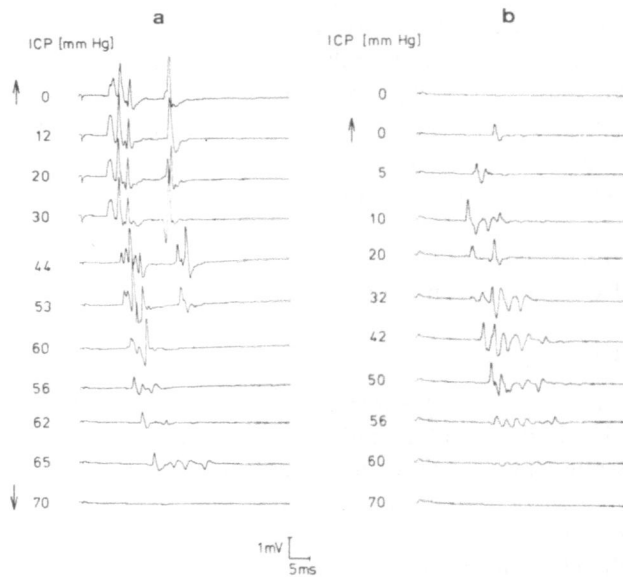

Fig. 20. Alterations of the blink reflex during repeated elevation of intracranial pressure. a) First elevation of pressure. b) Second elevation of pressure. ↑ = beginning of pressure elevation. ↓ = relief of pressure by deflating the balloon

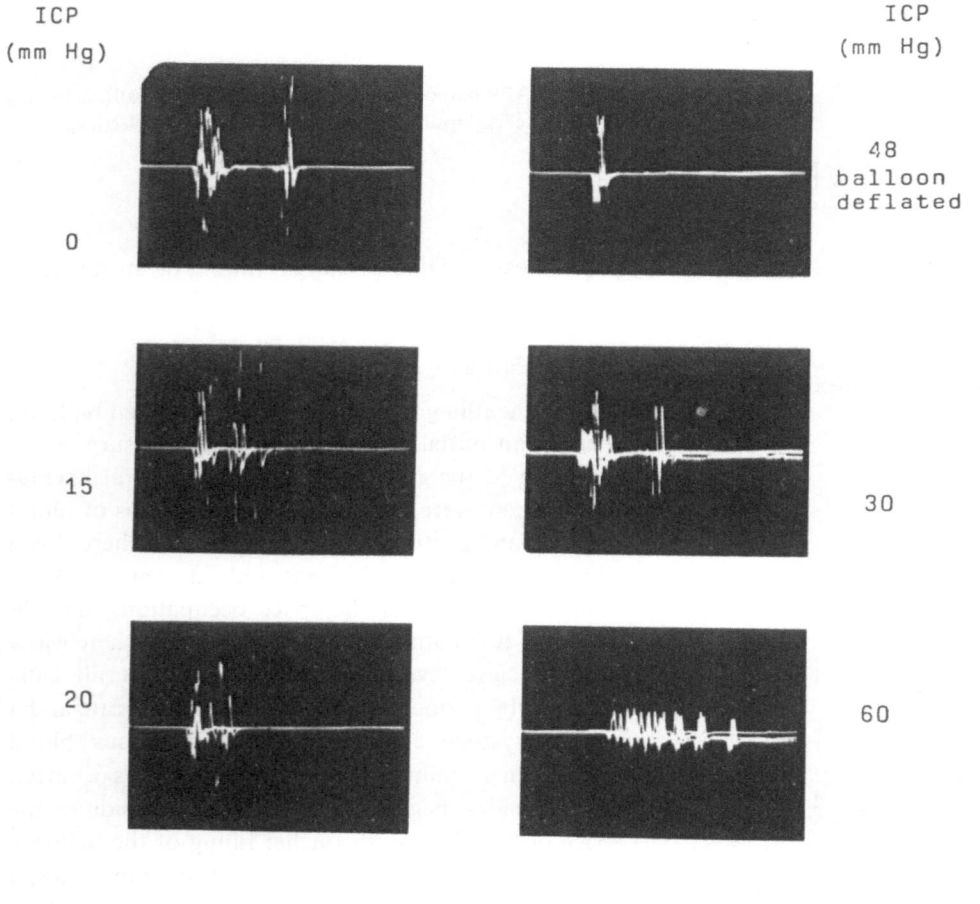

Fig. 21. Latency shortening and increase in amplitude of R_2 during the pressure elevation. R_2 recovered again after relief of the pressure

pathologically altered. During a second increase of intracranial pressure (Fig. 20 b), R_1 showed more pronounced alterations which indicate a prior damage during the first pressure elevation. After an initial shortening, the latency increased again, and the increase was especially pronounced before the final loss of the reflex. The duration of R_1 increased appreciably and was finally two to three times the initial value, whereas the amplitude rapidly decreased prefinally. R_1 disappeared shortly before occurrence of respiratory arrest and the fall of blood pressure. With a relief of pressure immediately after loss of the reflex response, R_2 recovered again (Fig. 21). The BR findings shown in Fig. 21 are a special feature to the extent that the shortening of latency and increase in amplitude only involve the late response.

Reanimation measures after the occurrence of respiratory arrest did not lead to a recovery of R_1 and R_2 despite immediate relief of intracranial pressure, although spontaneous breathing occurred once more and the blood pressure returned almost to its initial value.

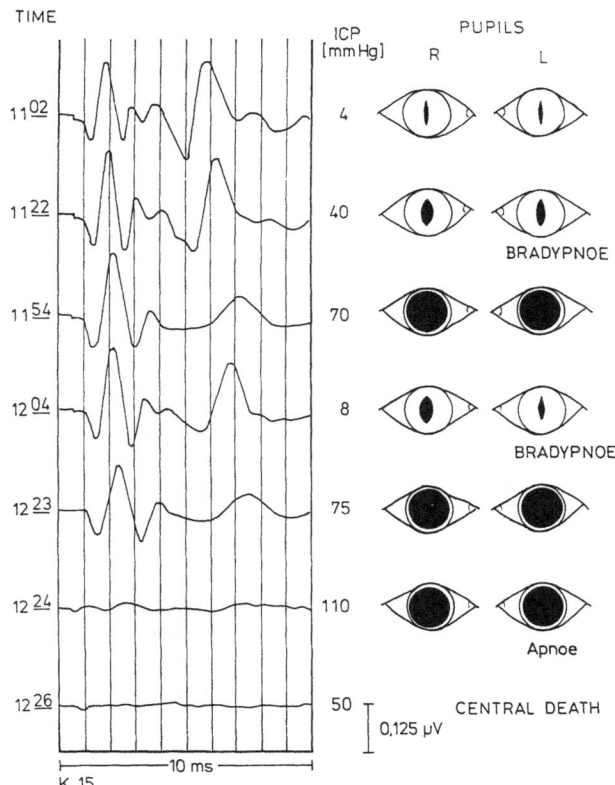

Fig. 22. BAEP during repeated elevation of intracranial pressure

4.4.2.3. Acoustic Evoked Potentials

During the rise of intracranial pressure, there was a marked latency delay and (at about 40 mm Hg) loss of wave V at an early stage. At the same time, irregularities of breathing and a decrease of the breathing rate was observed. With a further increase of ICP, the pupils became mydriatic and fixed. In some cats, wave III was lost at the same time contralateral to the space occupation (Fig. 22).

This finding indicates a functional disorder of pontine segments of the auditory tract which has already occurred at the time of the mesencephalic herniation. It must remain an open question as to why wave III is first suppressed contralaterally to the space occupation. A displacement of the brain stem contralateral to the haematoma is conceivable. On the other hand, it is known from animal experiments that wave III greatly depends on contralateral nuclear areas[1, 8], so that a loss of wave III contralaterally is also to be discussed owing to brain stem compression on the side of the haematoma. After relief of pressure at this time, there is normalization of the pupils and recovery of wave III, whereas wave V is no longer generated (Fig. 22). Accordingly, the suspicion can be expressed that the reversible pontine damage is less than the mesencephalic damage. On the other hand, the ques-

tion is raised as to whether the regression of the mydriasis in absence of recovery of wave V indicates a functional disorder of the oculomotor nerve rather than a lasting damage to collicular nuclear areas of the auditory tract (inferior colliculus). The BAEP alteration which occurred during the intracranial pressure elevation with prolonged latencies and interpeak latencies did not regress completely after the relief of pressure. During repeated pressure elevation, the alterations described already occurred at lower ICP values, possibly as a manifestation of prior damage during the first rise of pressure.

For evaluation of IPL alterations during the intracranial pressure elevation, the intraindividual values before filling the balloon were compared with the interpeak latencies in the stage of mydriasis: a significant prolongation (p less than 0.01) was present for I–III ($n = 12$), I–IV and I–V ($n = 11$). The IPL III–V was prolonged in eleven cats (p less than 0.05). This prolongation of the interpeak latencies in all segments investigated indicates a functional impairment of pontine, pontomesencephalic and mesencephalic segments of the auditory tract in the stage of incipient mesencephalic herniation.

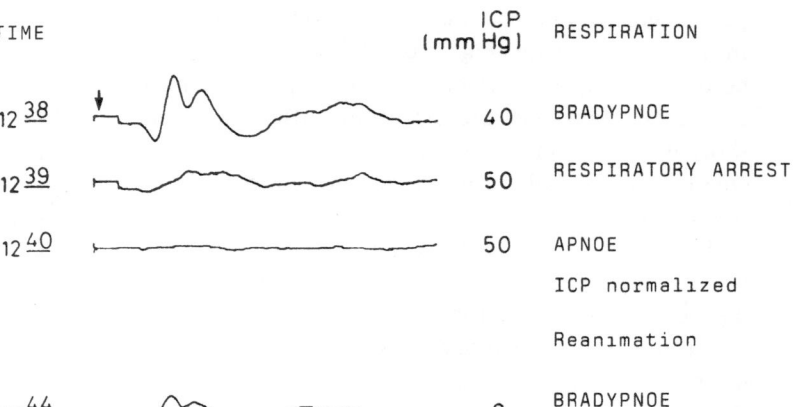

TIME		ICP (mm Hg)	RESPIRATION
12 38		40	BRADYPNOE
12 39		50	RESPIRATORY ARREST
12 40		50	APNOE
			ICP normalized
			Reanimation
12 44		0	BRADYPNOE

0,5 µV ⌐ 1ms

Fig. 23. BAEP in the final stage and in brain death after three elevations of intracranial pressure. After relief of pressure and reanimation, I and II are generated again. At the same time, insufficient spontaneous respiration starts

Prefinally, only waves I and II could be demonstrated. At this time, there was a bradypnea and dilated, nonreactive pupils. Within a very short time, there was respiratory arrest. The BAEP did not recover after the relief of pressure. On the other hand, waves I and II were generated again when reanimation measures were carried out in addition (Fig. 23).

In seven out of 12 cats, there was an initial increase in amplitude of individual potential peaks during the increase of intracranial pressure. As a rule, wave IV showed an initial rise (not statistically significant). For waves III, II and I, the increase in amplitude proved to be significant (Fig. 24). Wave V did not show any rise of amplitude.

The point of deflection for the amplitudes of the individual potential components from the ascending to the descending phase was on average at the following values of brain pressure: For wave IV at 10–20 mm Hg, for wave III at 30–40 mm Hg, for wave II at 50–60 mm Hg and for wave I at 40–60 mm Hg.

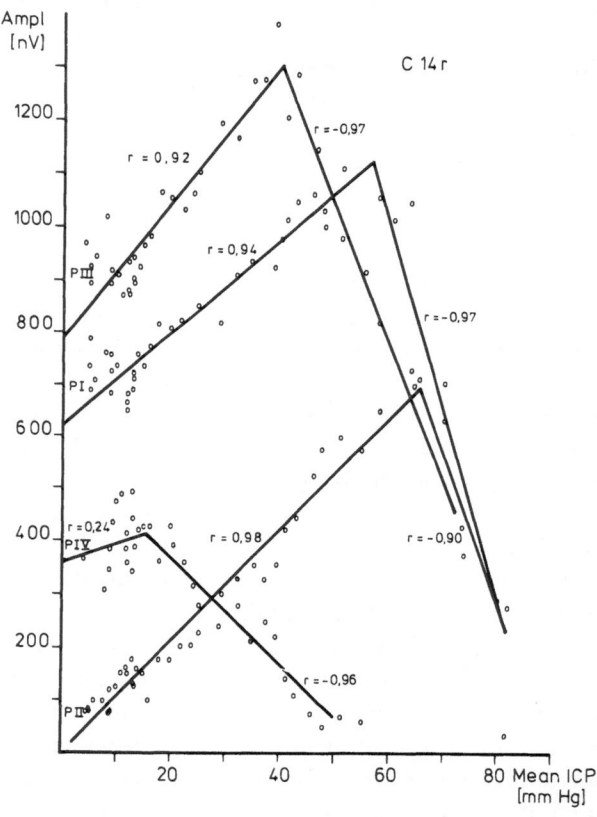

Fig. 24. Amplitude behaviour of individual potential components of BAEP in relation to the intracranial pressure. r = correlation coefficient

4.4.2.4. Pathomorphological and Histopathological Findings*

4.4.2.4.1. Macroscopic Findings

All brains showed a pronounced compression of the cortex in accordance with the epidural space occupa-

tion. The compression was located fronto-temporally in 16 cases, and in six cats it extended somewhat to parietal and more to temporal in five further cats. In the vicinity of the zone of compression due to the balloon, pronounced hyperemia was found with extravasation

---

* We are grateful to Privat-Dozent Dr. Z. M. Rap for provision of the histopathological and electron microscopic findings.

of blood in the subarachnoid space and the cortex with escape of Evans blue. Some cats showed indications of disturbances of the blood-brain barrier over the left hemisphere corresponding to the location of the ICP transducer (Fig. 25).

A herniation of the uncus and the splenium of the corpus callosum, the cingulate gyrus and the cerebellar tonsils were found as signs of manifest displacements (Fig. 26). Sagittal sections through the corpus callosum, the basal ganglia and the brain stem displayed vascular lesions in the specified structures involved in the herniation; these lesions were macroscopically recognizable owing to a blue discoloration and were more pronounced in all cases on the side of the epidural space occupation. A primary ischaemic zone in front of and behind the tentorium was found in all animals (Fig. 26 c). The zone of ischaemia situated in front of the tentorium extended from the cingulate gyrus through the splenium of the corpus callosum and the massa intermedia of the thalamus to the mammillary bodies of the hypothalamus.

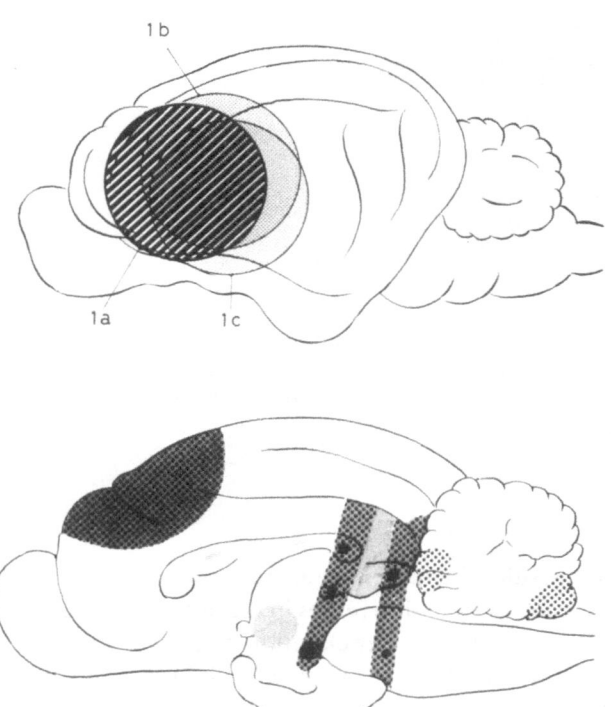

Fig. 25. Top: compression zones of the cortex after balloon compression extending to frontotemporal (*1 a*), parietal (*1 b*) and temporal (*1 c*). Bottom: macroscopically discernible ischaemia in front of and behind the tentorium recognizable in the sagittal section

The zone traversed the brain stem at the level of the mesodiencephalic transition. The second zone of ischaemia traversed the mesencephalon from the lamina quadrigemina (superior and inferior colliculus) through the pedunculus along the mesencephalopontine transition to basal parts of the pons. An escape of Evans blue into the cerebellar cortex and at the site of the caudally displaced lower corpora quadrigemina as well as the cerebellar tonsils was found in some cases. In all brains, the structures situated in front of the tentorium were damaged to a greater extent than those behind the tentorium. The ischaemic zones with BBB disorders in front of the tentorium showed secondary haemorrhages in 15 cats; these occurred especially frequently in the splenium of the corpus callosum, thalamus and posterior parts of the hypothalamus as well as in the region of the mamillary bodies. Macroscopically visible haemorrhages in the primary ischaemic zones behind the tentorium were found in five cats.

4.4.2.4.2. Fluorescence Microscopy Findings

In fluorescence microscopy, there was a diffuse red fluorescence in the regions in front of and behind the tentorium which already showed a blue discoloration macroscopically. The albumin-Evans blue complex could be demonstrated in all nervous structures of the ischaemic zones described above (Fig. 27 a). In the regions in front of (ventrolateral parts of the thalamus, median parts of the middle and posterior hypothalamus), between (subthalamus, red nucleus, lateral geniculate body, corticospinal projecting fibers) and behind (all parts of the pons and the medulla oblongata) the zones of ischaemia, the albumin-Evans blue complex was found selectively in neurons, their fibers and sometimes in glia cells (Fig. 27 b, c).

4.4.2.4.3. Light Microscopic Investigations

The structures situated in front of and behind the tentorium showed a congestion of capillaries, arterioles and venules as well as perivascular diapedesis haemorrhages in the regions of the ischaemic zones. Occasionally, large confluent haemorrhages into the posterior parts of the thalamus and mostly in the region of the mamillary bodies of the hypothalamus were found. Atrophic nerve cells with PAS-positive substances were demonstrated in the pons and medulla oblongata and in a few cases also in the cerebellar cortex as well as in particular in the motor nucleus areas of the IIIrd, IVth, Vth and VIIth cranial nerves. Nerve cell bundles were pushed apart, the perivascular spaces were widened and

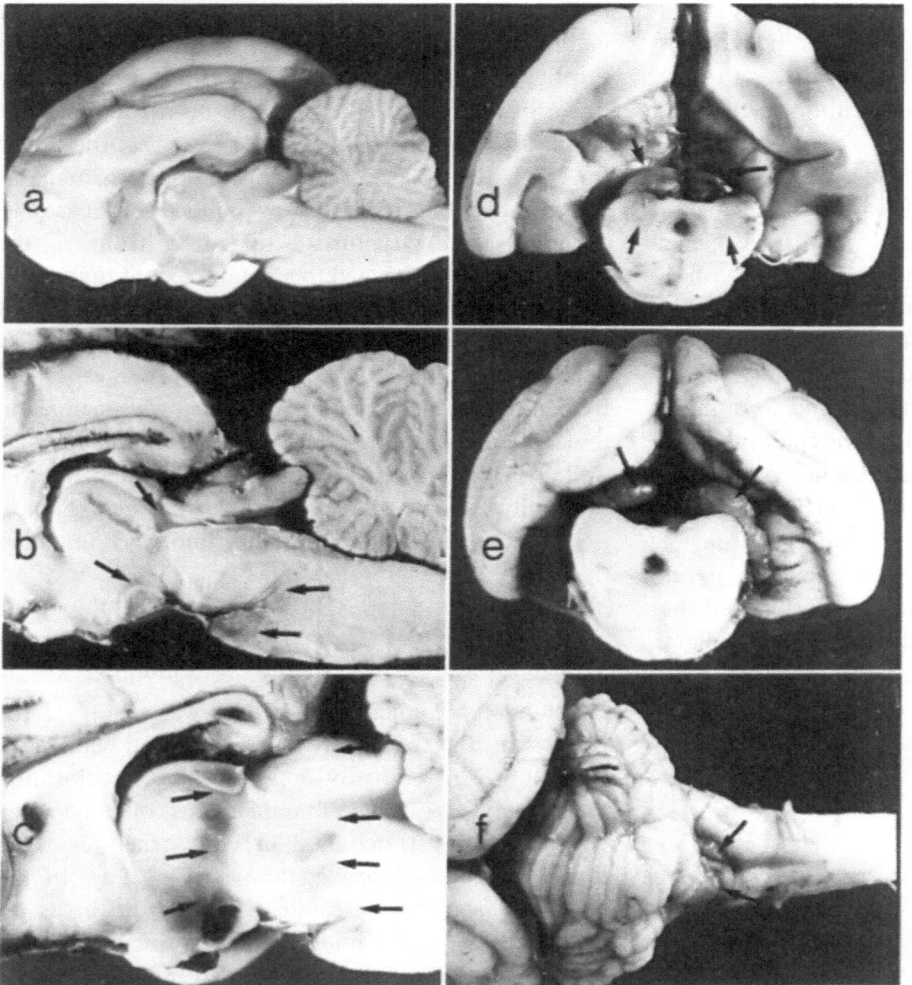

Fig. 26. Morphological findings in the cat brain after acute intracranial pressure elevation. a) Cat without Evans blue, no detection of structural alterations in the diencephalon and brain stem × 1.5. b) Primary ischaemic lesions with disturbance of the blood-brain barrier (BBB) in the diencephalon and mesencephalon. Evans blue. × 3. c) Primary ischaemic zones with secondary haemorrhages. Evans blue. × 6. d) BBB disorder in the cingulate gyrus, superior and inferior colliculus. × 2. e) Herniation of the lingual gyrus and cingulate gyrus as well as the lateral geniculate body with BBB disturbance. × 2. f) Herniation of the cerebellar tonsil with BBB disturbance. × 3

oligodendrocytes were swollen as a manifestation of the incipient brain oedema.

4.4.2.4.4. Electron Microscopic Investigations

In particular alterations of the vessels in the ischaemic zones were demonstrated in electron microscopy. In these areas, there was acute swelling of the endothelial cells of capillaries, precapillary and postcapillary venules with aggregates of blood cells. Besides this, swollen perivascular glial fibers and a disintegration of the myelin sheath were found.

4.4.3. Discussion

The results obtained in the acute pressure experiment in the cats revealed differences in the initial and late stage of intracranial pressure elevation: Initially, there was an increase in amplitude in various BAEP components and both blink reflex responses which initially displayed a shortening of latency. With a further rise of pressure, the latencies of BAEP and the reflex responses increased. Associated in time with clinical signs of mesencephalic herniation, waves V and R_2 were suppressed, followed by a loss of further BAEP components in the rostrocaudal direction. Before onset of respiratory arrest, R_1 and (with manifest respiratory paralysis) wave II disappeared, followed by wave I.

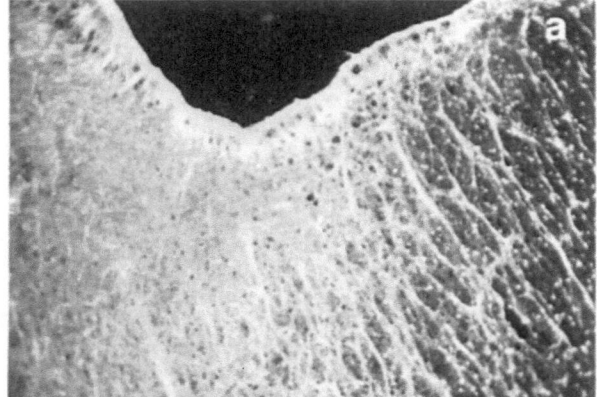

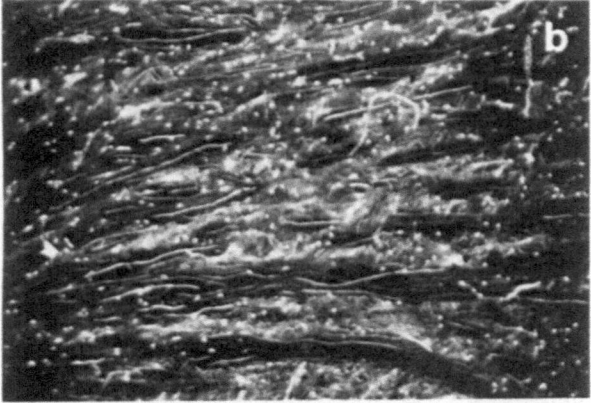

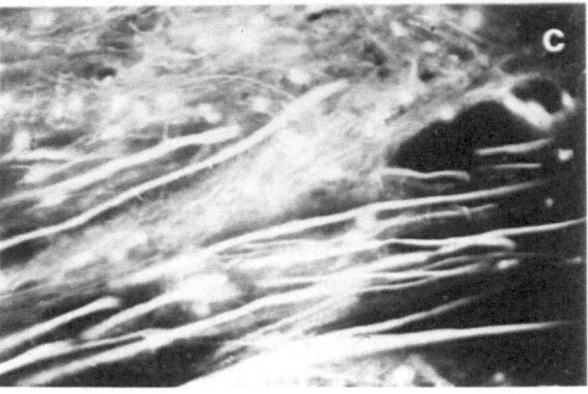

Fig. 27. Red fluorescence of Evans blue-albumin complex in nerve fibers, neurons and glial cells at the boundary between the mesencephalon and the pons. × 25

With clinical brain death, BR and BAEP were no longer recovered. Immediate reanimation with recommencement of spontaneous breathing did not affect the blink reflex. Waves I and II were generated again as the sole reaction. The pontine and mesencephalic parts of the nucleus and tract systems remained without reaction and document the irreversible damage. It results from the morphologically confirmed manifestations of incarceration with herniation in the region of the diencephalon, midbrain, pons and medulla oblongata and

secondary ischaemic alterations. Two zones with disorders of the blood-brain barrier parallel to the tentorium were demonstrated in the region of the mesodiencephalic and pontomesencephalic transition. In connection with the BR and BAEP alterations described, it appears justified to assume that the disorder of the barrier first of all affects the zone in front of the tentorium, and later the zone behind the tentorium. The finding of a potential suppression in the rostrocaudal direction found during the elevation of intracranial pressure corresponds to our own clinical findings and the experimental results of other authors[105]. On the other hand, the initial increase in amplitude of BAEP found during the raised pressure in the cat was unexpected.

An initial increase in activity in the mesencephalic and pontine reticular formation during the increase in intracranial pressure in cats is known[49, 50]. During an increase of intracranial pressure in cats, there was an increase in activity in the red nucleus and the giagantocellular nucleus, whereas activity decreased in the lateral geniculate nucleus situated further rostrally as soon as the brain pressure rose. Two factors were discussed to explain the initial increase in activity: on the one hand, metabolism with neurotransmitters acting on neurons in the nuclear areas in the brain stem, and on the other hand local or regional cerebral blood flow[48]. Investigations of our own study group in the same animal model[170] indicate an influence of regional blood flow in the brain (Fig. 28).

Fig. 28 shows that regional blood flow in the brain is maintained with falling cerebral perfusion pressure (to be read from the right to the left in the figures), or indeed increases slightly (pons and rostral medulla oblongata).

On the one hand the inhibitory and on the other hand activity-enhancing influence of neurotransmitters in neuronal regulation in the course of the auditory tract in the brain stem has been known for a long time[29, 158, 159].

Administration of 5-hydroxytryptamine (5-HT), noradrenaline (NA) and dopamine (DA) (these already occur under normal conditions in the medial geniculate body) to strychnine-sensitive neurons in the medial geniculate body of the cat led to a decrease of discharging units[159]. Intraperitoneally applied reserpine, which brings about an initial release of presynaptic 5-hydroxytryptamine and another monoamines, led to a biphasic effect on the amplitude behaviour of wave III and wave IV of the brain stem acoustic evoked potential. Initially, there was a reduction of amplitude in the first 30 to 60 minutes and an increase in

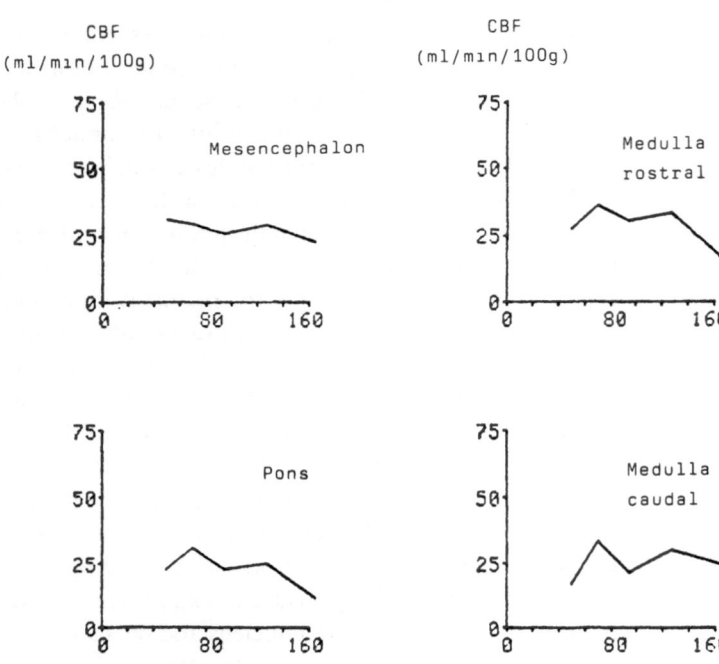

Fig. 28. rCBF in various regions of the brain stem depending on the cerebral perfusion pressure

amplitude after 90 to 120 minutes[6]. At a low concentration, eserine brought about a shortening of latency and amplitude reduction of wave III and wave IV. At a high concentration, there was a latency delay and an amplitude increase of waves II, III and IV. This biphasic effect was more pronounced with carbachol, whereas atropine did not show any action on latency behavior, but led to a reduction of BAEP amplitude[5]. Other authors have demonstrated an increased secretion of ACTH and primarily of catecholamines into the bloodstream and into the CSF during intracranial pressure elevation[121]. On the basis of our own experimental results and in connection with the described alterations of neurotransmitters, regional brain stem blood flow and the initial increase in activity in various regions of the brain stem at the beginning of the elevation of intracranial pressure, it is suspected that the rise in intracranial pressure initially leads to activation of synaptic relay stations in the brain stem. This results in an increase in their amplifier function, which is indirectly manifested in an increase in amplitude of individual BAEP components. The increase in amplitude and shortening of the latency of both responses of the blink reflex up to ICP values around 35–45 mm Hg is consistent with these findings. On the other hand, there was an increase in amplitude and shortening in latency of R_1 after precollicular postmamillary decerebration in cats[63]. This result might be due to a loss of inhibitory neurons. An analogous amplitude and latency behaviour of the R_1 response was found after total cerebellectomy. The authors themselves discuss the possibilities of an abolition of possible inhibitory influences on the part of the cerebellum.

5. Summary

In the present paper, the function of the brain stem in patients with brain stem lesions of various aetiology is investigated with electrophysiological methods. The clinical observations are supplemented by experimental investigations on cats, in which the blink reflex and the early acoustic evoked potentials were registered during the acute elevation of intracranial pressure. The findings in patients with circumscribed space-occupying lesions in the posterior fossa document that the registration of the BR and the BAEP have a functional diagnostic significance above and beyond the neurological and radiological investigation. In the case of the cerebellar space occupations, specific alterations could not be observed. On the contrary, the alterations of BR and BAEP indicate a general disturbance of brain stem function, possibly as a result of a general increase of intracranial pressure. In cerebellopontine angle tumours, both BR and BAEP showed specific alterations which were usually asymmetrical. The BR changes ipsilateral to the tumour are of major topodiagnostic significance, whereas the alterations of the contro-lateral potential are especially informative in the registration of BAEP. The alterations of BR and BAEP also allow an appraisal of the localization and extent of the lesion in primary space occupations in the brain stem: A pathological R_1 indicates a pontine lesion, whereas pathological R_2 responses are found in medullary and in oral pontine and mesencephalic lesions. In contrast to cerebellopontine angle tumours, the BAEP tends to show symmetrical alterations in primary brain stem lesions. The prolongations of interpeak latencies correspond to the brain stem segment concerned, and the same also applies to pathological amplitude reduction and deformations of individual potentials. In patients with localized brain stem damage, the reflex pathway of R_2 is discussed on the basis of the BR findings. In contrast to the view held up to now that only structures situated caudal of the facial nucleus area are responsible for the genesis of the R_2 response, it is assumed on the basis of our own observations that pontomesencephalic structures rostral to the facial nuclei are also important for the genesis of R_2. Registration of BR and BAEP in patients with acute diffuse brain stem damage shows that both methods have a high diagnostic and prognostic value. Isolated damage and combined brain stem lesion can be demonstrated and the course can be followed up.

Normalization of pathological findings reflects clinical recovery, and conversely a secondary deterioration indicates the presence of complications. Alterations of both components of the blink reflex are appropriate to establish the lesion level and disorders in the brain stem which are more pronounced on one side. Moreover, in the individual case they allow differentiation between primary and secondary damage.

As against the intraindividual and interindividual time stability of the individual potential components in healthy subjects, brain stem acoustic evoked potentials show an appreciable instability of latency in acute midbrain syndrome. The interpeak latencies, which are regularly very prolonged for the medullopontine, pontomesencephalic and central conduction time indicate a functional disorder both of the oral and of the caudal brain stem. Further typical features are pronounced alterations of individual potential components in the form of a flattening, broadening and desynchronization as well as a reduction of amplitude and a decrease of the amplitude ratios III/I, V/III and V/I. The development of an acute midbrain syndrome via the bulbar syndrome to brain death is characterized by loss of the potential components in the rostrocaudal direction. In view of all the alterations specified, a graduation of BAEP into six degrees of severity appears to be reasonable. With application of these graduations, a single registration already allows certain prognostic inferences.

The BR and BAEP results obtained in the acute pressure study in the animal model reveal differences between the early and late stage of intracranial pressure elevation. R_2 is suppressed simultaneously with the clinical signs of mesencephalic herniation, whereas R_1 only disappears just before onset of respiratory arrest. This finding likewise indicates the significance of ponto-mesencephalic structures for the genesis of R_2 which has already been discussed. In the early stage of the pressure elevation, there is an increase in amplitude of the two reflex responses and a shortening of both latencies; an increase in the amplitudes of the waves I, II, III and occasionally IV was also observed in the same phase. With a further increase of pressure, there was a progressive decrease in amplitude and a progressive increase in the latencies of the reflex responses and the BAEP waves. R_2 and wave V were suppressed at approximately the same time, whereas R_1 and wave III disappeared before the onset of respiratory arrest, followed by waves II and I. After onset of clinical brain death, BR and BAEP could no longer be demonstrated. Immediate relief of pressure in connection with reanimation measures did not affect the blink reflex, whereas the waves I and II were generated again, accompanied by a recommencement of spontaneous breathing.

The morphological and histological investigation of the cat brains led to demonstration of two zones of ischaemia in parallel to the tentorium along the meso-diencephalic and pontomesencephalic transition region. In connection with the BR and BAEP alterations described, it is probable that the disturbance of the barrier first of all affects the area in front of the tentorium and only later affects the zone situated behind it. The finding of an increase in amplitude of the two blink reflex responses and individual BAEP components as well as a shortening of R_1 and R_2 latency at the beginning of the rise of intracranial pressure was unexpected and has not been described so far in the literature. These results which are surprising at first glance are discussed in the context of communications from the literature.

References

1. Achor LJ, Starr A (1980) Auditory brain stem responses in the cat. I. Intracranial and extracranial recordings. EEG Clin Neurophysiol 48: 154–173

2. Achor LJ, Starr A (1980) Auditory brain stem responses in the cat. II. Effects of lesions. EEG Clin Neurophysiol 48: 174–190

3. Allen AR, Starr (1978) Auditory brain stem potentials in monkey (M. Mulatta) and men. EEG Clin Neurophysiol 45: 53–63

4. Amadeo M, Shagass C (1973) Brief latency click evoked potentials during waking and sleep in man. Psychophysiol 10: 244–250

5. Bhargava VK, Salamy A et al (1976) Effect of cholinergic drugs on the auditory evoked responses from brain stem to auditory cortex. Abstr Neuroscience 2: 14

6. Bhargava VK, McKean CM (1977) Role of 5-hydroxy-tryptamine in the modulation of acoustic brain stem (far-field) potentials. Neuropharmacology 16: 447–449

7. Brainin M, Binder H et al (1983) Prognostic value of the late blink-reflex in tegmental brain stem haemorrhage. Neuropsych Clin 2: 57–63

8. Buchwald JS, Huang CM (1975) Origins of the far-field acoustic response in the cat. Science 189: 382–384

9. Buonaguidi R, Rossi B et al (1979) Blink reflexes in severe traumatic coma. J Neurol Neurosurg Psychiat 42: 470–474

10. Caird D, Sontheimer D, Klinke R (1985) Intra- and extracranially recorded auditory evoked potentials in the cat. I. Source location and binaural interaction. EEG Clin Neurophysiol 61: 50–60

11. Cashman MZ, Rossman RN (1983) Diagnostic features of the auditory brain stem response in identifying cerebellopontine angle tumours. Scand Audiol 12: 35–41

12. Chiappa KH (1982) Physiologic localization using evoked responses: pattern shift visual, brain stem auditory and short latency somatosensory. In: Thompson RA et al (eds) New perspectives in cerebral localization. Raven Press, New York

13. Chiappa KH, Gladstone KJ, Young RR (1979) Brain stem auditory evoked responses: studies of variations in 50 normal human subjects. Arch Neurol 36: 81–87

14. Chiappa KH, Ropper AH (1982 a) Evoked potentials in clinical medicine (first of two parts). New England J Med 19: 1140–1150

15. Chiappa KH, Ropper AH (1982 b) Evoked potentials in clinical medicine (second of two parts). New England J Med 20: 1205–1211

16. Clay SA, Ramseyer JC (1977) The orbicularis oculi reflex: pathologic studies in childhood. Neurology 27: 892–895

17. Clemis JD, McGee T (1979) Brain stem electric response audiometry in the differential diagnosis of acoustic tumours. The Laryngoscope 89: 31–42

18. Clemis JD, Mitchel C (1977) Electrocochleography and brain stem responses used in the diagnosis of acoustic tumours. J Otolaryng 6: 447–459

19. Coats AC, Martin JL (1977) Human auditory nerve action potentials and brain stem evoked responses. Arch Otolaryngol 103: 605–622

20. Courjon J, Maugière F, Revol M (1982) Clinical applications of evoked potentials in neurology. Advances in neurology 32. Raven Press, New York

21. Courville J (1966) The nucleus of the facial nerve: the relation between cellular groups and peripheral branches of the nerve. Brain Res 1: 338–354

22. Csécsei G (1979) Facial afferent fibers in the blink reflex of man. Brain Res 161: 347–350

23. Csécsei G (1980) Alteration of blink reflexes in hemispheral lesions. Electromyogr Clin Neurophysiol 20: 141–152

24. Csécsei G (1982) Facial reflexes of short latency. Electromyogr Clin Neurophysiol 22: 39–44

25. Csécsei G, Klug N (1986) Der Einfluß umschriebener medullärer und pontomesencephaler Schädigungen auf die Spätantwort des Blinkreflexes. In: Lowitzsch K (ed) Hirnstammreflexe. Methodik und klinische Anwendung. G Thieme, Stuttgart New York, pp 132–137

26. Csécsei G, Martini E (1981) Prognostic value of the blink reflex in comatouse patients. Eur Neurol 20: 473–477

27. Csécsei G, Klug N, Rap Z (1983) Effect of increased intracranial pressure on blink reflex in cats. Acta Neurochir (Wien) 68: 85–92

28. Curio G, Oppel F, Scherg M (1987) Peripheral origin of BAEP wave II in a case with unilateral pontine pathology: a comparison of intracranial and scalp recordings. EEG Clin Neurophysiol 66: 29–33

29. Curtis DR, Koizumi K (1961) Chemical transmitter substances in brain stem of cat. J Neurophysiol 24: 80–90

30. Davis SL, Aminoff MJ, Berg BO (1985) Brain stem auditory evoked potentials in children with brain stem or cerebellar dysfunction. Arch Neurol 42: 156–160

31. Dawson GD (1951) A summation technique for detecting small signals in a large irregular background. J Physiol 2: 115

32. Dawson GD (1954) A summation technique for the detection of small evoked potentials. EEG Clin Neurophysiol 6: 65–84

33. Dehen H, Willer JC et al (1976) Blink reflex in hemiplegia. EEG Clin Neurophysiol 40: 393–400

34. Dengler R, Struppler R (1981) Beurteilung der Lokalisation und Ausdehnung von Hirnstammaffektionen mit Hilfe des Orbicularis-oculi-Reflexes. Z EEG-EMG 12: 50–55

35. Dengler R, Kossev A et al (1982) Quantitative analysis of blink reflexes in patients with hemiplegic disorders. EEG Clin Neurophysiol 53: 513–524

36. Ebner A, Scherg M, Dietl H (1980) Das akustisch evozierte Hirnstammpotential in der klinisch-neurologischen Anwendung. Z EEG-EMG 11: 205–210

37. Edvinsson L, Owman C et al (1971) Brain concentrations of dopamine, noradrenaline, 5-hydroxytryptamine and homovanillic acid during intracranial hypertension following traumatic brain injury in rabbit. Acta Neurol Scand 47: 458–463

38. Eggermont JJ (1984) Use of electrocochleography and brain stem auditory evoked potentials in the diagnosis of cerebellopontine angle pathology. Adv Otorhinolaryngol 34: 47–56

39. Eggermont JJ, Don M, Brackmann DE (1980) Electrocochleography and auditory brain stem electric responses in patients with pontine angle tumours. Ann Otol Rhinol Laryngol 89: 1–19

40. Eisen A, Danon J (1974) The orbicularis oculi reflex in acoustic neuromas: a clinical and electrodiagnostic evaluation. Neurology 4: 306–311

41. Emmert H, Skiba N, Flügel KA (1987) A comparison of BAEP and NMR findings in patients with brain stem lesions. Clin Neurol Neurosurg [Suppl 1] 89, 2: 76

42. Faught E, Oh SJ (1985) Brain stem auditory evoked responses in brain stem infarction. Stroke 16: 701–705

43. Feinmesser P, Goldstein KJ, Sohmer H (1982) The relationship between intracranial pressure and auditory brain stem evoked response, diagnostic and prognostic implications. Fifth Internat Symp on Intracranial Pressure, Tokyo, 30.5.–2.6.1982

44. Fisher CM (1967) Some neuro-ophthalmological observations. J Neurol Neurosurg Psychiat 30: 383–392

45. Fisher CM (1969) The neurological examination of the comatose patients. Acta Neurol Scand [Suppl 36] 45: 1–56

46. Fisher MA, Shahani BT, Young RR (1979) Assessing segmental excitability after acute rostral lesions: the blink reflex. Neurology (Minneap) 29: 45–50

47. Garg BP, Markand ON, Bustion PF (1982) Brain stem auditory evoked responses in hereditary motorsensory neuropathy: site of origin of wave II. Neurology 32: 1017–1019

48. George B (1980) Neurophysiological effects of experimental intracranial hypertension on three different structures of the brain stem in the cat. Rostrocaudal deterioration. Acta Neurochir (Wien) 55: 71–83

49. George B, Benoit O (1977) Activité réticulaire et pression intracranienne. Hypertension intracranienne aique et chronique. Acta Neurochir (Wien) 38: 195–209

50. George B, Levante A et al (1977) Tentative d'explication des décompensations d'hydrocéphalie par des modifications de l'activité réticulaire. Neurochirurgie 23: 37–46

51. Gilroy J, Lynn GE et al (1977) Auditory evoked brain stem potentials in a case of "locked-in"-syndrome. Arch Neurol 34: 492–495

52. Greenberg RP, Mayer DJ et al (1977 a) Evaluation of brain function in severe human head trauma with multimodality evoked potentials. Part 1: Evoked brain-injury potentials, methods and analysis. J Neurosurg 47: 150–162

53. Greenberg RP, Becker DP et al (1977 b) Evaluation of brain function in severe human head trauma with multimodality evoked potentials. Part 2: Localization of brain dysfunction and correlation with posttraumatic neurological conditions. J Neurosurg 47: 163–177

54. Greenberg RP, Ducker TB (1982) Evoked potentials in the clinical neurosciences. J Neurosurg 56: 1–18

55. Greenberg RP, Becker JD (1980) Early prognosis after severe human head injury utilizing multimodality evoked potentials. Acta Neurochir (Wien) [Suppl 28]: 50–51

56. Greenberg RP, Newlon PG (1981) Prognostic implications of early multimodality evoked potentials in severely head-injured patients. J Neurosurg 55: 227–236

57. Greenberg RP, Stablein DM, Becker DP (1981) Noninvasive localization of brain stem lesions in the cat with multimodality evoked potentials. J Neurosurg 54: 740–750

58. Hacke W, Schaff C, Zeumer H (1983) Der Orbicularis-oculi-Reflex bei computertomographisch verifizierten Läsionen der hinteren Schädelgrube. Fortschr Neurol Psych 51: 313–324

59. Hall JW, Tucker DA (1986) Sensory evoked responses in the intensive care unit. Ear Hear 7: 220–232

60. Hall JW III, Huang-Fu M, Gennarelli TA (1983) Auditory function in acute severe head injury. Laryngoscope 92: 883–890

61. Hall JW, Mackey-Hargadine J (1984) Auditory evoked responses in severe head injury. Seminars in Hearing 5: 313–336

62. Hammond EJ, Wilder BJ et al (1985) Auditory brain stem potentials with unilateral pontine haemorrhage. Arch Neurol 42: 767–768

63. Hiraoka M, Shimamura M (1977) Neural mechanisms of the corneal blinking reflex in cats. Brain Res 125: 265–275

64. Hoffmann O, Kurzaj E et al (1986) Blood flow in the brain stem during acute cerebral compression. In: Krieglstein J (ed) Pharmacology of cerebral ischaemia. Elsevier Science Publ, Amsterdam New York Oxford, pp 256–264

65. Holstege G, Kuypers HGJM (1977) Propriobulbar fibre connections to the trigeminal facial and hypoglossal motor nuclei.

I. An anterograde degeneration study in the cat. Brain 100: 239–264

66. Holstege G, Kuypers HGJM, Dekker JJ (1977) The organisation of the bulbar fibre connections to the trigeminal, facial and hypoglossal motor nuclei. II. An autoradiographic tracing study in cat. Brain 100: 265–286

67. Hopf HC (1986) Anatomische und physiologische Grundlagen der Hirnstammdiagnostik — Wertung der Befunde in der klinischen Praxis. In: Lowitzsch K (ed) Hirnstammreflexe. Methodik und klinische Anwendung. G Thieme, Stuttgart New York, pp 166–177

68. Hopf HC (1985) Die Generatoren der AEP-Wellen II-III. Akt Neurol 12: 58–61

69. Jewett DL (1970) Volume conducted potentials in response to auditory stimuli as detected by averaging in the cat. EEG Clin Neurophysiol 28: 609–618

70. Jewett DL, Romano MN, Williston JS (1970) Human auditory evoked potentials: possible brain stem components detected on the scalp. Sciences 167: 1517–1581

71. Jewett DL, Williston JS (1971) Auditory evoked far fields averaged from the scalp of humans. Brain 94: 681–696

72. Kaplan PE, Bonis G et al (1977) The use of the blink reflex in evaluating the patient with stroke and communication disorders. EMG Clin Neurophysiol 17: 333–338

73. Kimura J (1970) Alteration of the orbicularis oculi reflex by pontine lesions. Study in multiple sclerosis. Arch Neurol Chir 22: 156–161

74. Kimura J (1971) Electrodiagnostic study of brain stem strokes. Stroke 2: 576–586

75. Kimura J, Lyon LW (1972) Orbicularis oculi reflex in the Wallenberg snydrome: alteration of the late reflex by lesions of the spinal tract and nucleus of the trigeminal nerve. J Neurol Neurosurg Psychiat 35: 228–233

76. Kimura J, Lyon LW (1973) Alteration of orbicularis oculi reflex by posterior fossa tumors. J Neurosurg 38: 10–16

77. Kimura J (1975) Electrically elicited blink reflex in diagnosis of multiple sclerosis. Review of 260 patients over a seven-year period. Brain 98: 413–426

78. Kjaer M (1979) Evaluation and graduation of brain stem auditory evoked potentials in patients with neurological diseases. Acta Neurol Scand 66: 231–242

79. Kjaer M (1980 a) Localizing brain stem lesions with brain stem auditory evoked potentials. Acta Neurol Scand 61: 265–274

80. Klug N (1982) Frühe akustisch evozierte Potentiale (FAEP) im Decerebrations- und Bulbärhirnsyndrom, sowie beim zentralen Tod. In: Struppler A (ed) Electrophysiologische Diagnostik in der Neurologie. G Thieme, Stuttgart New York

81. Klug N (1982) Brain stem auditory evoked potentials in syndromes of decerebration, bulbar syndrome and in central death. J Neurol 227: 219–228

82. Klug N (1983) Funktionsuntersuchungen des Hirnstamms im akuten Mittelhirnsyndrom unter Berücksichtigung vegetativer Meßgrößen während der Dezerebration. Habil Schrift, Gießen

83. Klug N, Csécsei G (1985) Evoked potentials and brain stem reflexes. New diagnostic devices in neurosurgery. Neurosurg Rev, pp 63–84

84. Klug N, Csécsei G (1986) Blinkreflex und frühe akustisch evozierte Potentiale bei Kleinhirnbrückenwinkeltumoren. In: Lowitzsch K (ed) Hirnstammreflexe. Methodik und klinische Anwendung. G Thieme, Stuttgart New York, pp 183–194

85. Kountouris D, Fritze J et al (1984) Blink reflex and trigeminal

nerve somatosensory evoked potentials: essentials in vascular brain stem diseases. Monogr Neural Sci 11: 228

86. Kraus N. Ozdamar O *et al* (1984) Auditory brain stem responses in hydrocephalic patients. EEG Clin Neurophysiol 59: 310–317

87. Kugelberg E (1952) Facial reflexes. Brain 75: 227–332

88. Lacquaniti F, Benna P *et al* (1979) Brain stem auditory evoked potentials and blink reflex in quiescent multiple sclerosis. EEG Clin Neurophysiol 47: 607–610

89. Lev A, Sohmer H (1972) Sources of averaged neural responses recording in animal and human subjects during cochlear audiometry (electro-cochleogram). Graefes Arch Clin Exp Ophthalmol 201: 79–90

90. Lyon LW, Kimura J, McCormick WF (1972) Orbicularis oculi reflex in coma: clinical, electrophysiological and pathological correlations. J Neurol Neurosurg Psychiat 35: 582–588

91. Lyon LW, Van Allen MW (1972) Alteration of the orbicularis oculi reflex by acoustic neuroma. Arch Otolaryng 95: 100–103

92. Mackey-Hargadine JR, Hall JW (1985) Sensory evoked responses in head injury. Central Nervous System Trauma 2: 187–206

93. Maurer K, Leitner H *et al* (1979) Frühe akustisch evozierte Potentiale (FAEP). Eine geeignete Screeningmethode zur Früherfassung des Akustikusneurinoms. Akt Neurol 6: 71–80

94. Maurer K, Leitner H, Schäfer E (1980) Detection and localization of brain stem lesions with auditory brain stem potentials. In: Barber C (ed) Evoked potentials. MTP Press, Lanchester, pp 391–398

95. Maurer K, Schäfer E *et al* (1980) The location by early auditory evoked potentials (EAEP) of acoustic nerve and brain stem demyelination in multiple sclerosis (MS). J Neurol 223: 43–58

96. Maurer K, Marneros A *et al* (1979) Early auditory evoked potentials (EAEP) in vertebral basilar insufficiency. Arch Psych Nervenkr 227: 367–476

97. McNealy DE, Plum E (1962) Brain stem dysfunction with supratentorial mass lesions. Arch Neurol 7: 10–32

98. Mehta AJ, Seshia SS (1976) Orbicularis oculi reflex in brain death. J Neurol Neurosurg Psychiat 39: 784–787

99. Mills JA, Ryals BM (1985) The effects of reduced cerebro-vascular circulation on the auditory brain stem response (ABR). Ear Hear 6: 139–143

100. Minami T, Kurokawa T *et al* (1984) Primary brain stem haemorrhage in a child: usefullness of auditory brain stem response (ABR). Neuropediatrics 15: 99–101

101. Miwa S. Inagaki C *et al* (1982) The activities of noradrenergic and dopaminergic neuron systems in experimental hydrocephalus. J Neurosurg 57: 67–73

102. Musiek FE, Josey AF, Glasscock ME (1986) Auditory brain stem response in patients with acoustic neuromas. Wave presence and absence. Arch Otolaryng Head Neck Surg 112: 186–189

103. Musiek FE, Kibbe K (1986) Auditory brain stem response wave IV–V abnormalities from the ear opposite large cerebello-pontine lesions. Am J Otol 7: 253–257

104. Musiek FE, Kibbe-Michal K *et al* (1986) ABR results in patients with posterior fossa tumours and normal pure-tone hearing. Otolaryng Head Neck Surg 94: 568–573

105. Nagao S, Roccaforte P, Moody RA (1979) Acute intracranial hypertension and auditory brain stem responses. Part 1: Changes in the auditory brain stem and somatosensory evoked responses in intracranial hypertension in cats. J Neurosurg 51: 669–676

106. Nagao S, Sunami N *et al* (1982) Serial observation of brain stem function in acute intracranial hypertension by auditory brain stem responses. A clinical study. 5th Internat Symp on ICP, Tokyo, 30.5.—3.6.1982

107. Namerow NS, Etemadi A (1970) The orbicularis oculi reflex in multiple sclerosis. Neurology 20: 1200–1203

108. Nishimoto H, Tsubokawa T *et al* (1981) Evaluation of brain stem damage in severe head injury by far field acoustic response. Neurol Med Chir (Tokyo) 21: 1147–1152

109. Nodar RH, Kinney SE (1980) The contralateral effects of large tumours on brain stem auditory evoked potentials. The Laryngoscope 90: 1762–1768

110. Ogasawara K (1985) Neural pathways mediating the corneal blink reflex and Bell's phenomenon in the cat. Neurosci Res 2: 309–320

111. Ohresser M, Toupet M *et al* (1986) Les neurinomas a BERA normaux. Ann Otolaryng Chir Cervicofac 103: 215–221

112. Olbrich HM, Nau HE *et al* (1986) Evoked potential assessment of mental function during recovery from severe head injury. Surg Neurol 26: 112–118

113. Ongerboer de Visser BW, Goor C (1974) Electromyographic and reflex study in idiopathic and symptomatic trigeminal neuralgias: latency of the jaw and blink reflexes. J Neurol Neurosurg Psychiat 37: 1225–1230

114. Ongerboer de Visser BW, Melchelse K, Megens PHA (1977) Corneal reflex latency in trigeminal nerve lesions. Neurology 27: 1164–1167

115. Ongerboer de Visser BW, Kuypers HGJM (1978) Late blink reflex changes in lateral medullary lesions. An electrophysiological and neuroanatomical study of Wallenberg's syndrome. Brain 101: 285–294

116. Ongerboer de Visser BW, Moffie D (1979) Effects of brain stem and thalamic lesions on the corneal reflex. An electrophysiological and anatomical study. Brain 102: 595–608

117. Overend W (1896) Preliminary note on a new cranial reflex. Lancet 1: 619–621

118. Parker SW, Chiappa KH, Brooks EB (1980) Brain stem auditory responses (BAERS) in patients with acoustic neuromas and cerebellopontine angle (CPA) meningiomas. Neurology 30: 413–414

119. McPherson DL, Amelie R, Foltz E (1985) Auditory brain stem response in infant hydrocephalus. Child's Nerv Syst 1: 70–76

120. Raman PT, Reddy PK, Rao SV (1976) Orbicularis oculi reflex and facial muscle electromyography. J Neurosurg 44: 550–555

121. Rap ZM, Staszewska-Barczak J (1975) Adrenergic response and morphologic changes in the neurosecretory system and adrenal cortex during intracranial hypertension in cats. In: Kormyei, Tariska (eds) VIIth International Congress of Neuropathology, Budapest/Hungary 1.—7.9.1974. Excerpta Medica, Amsterdam, Akademiai Kiado Budapest, pp 623–626

122. Riffel B, Stoehr M *et al* (1986) Evozierte Potentiale. Indikationen und diagnostische Bedeutung. Wien Med Wschr 136: 323–328

123. Rossi B, Sartucci F *et al* (1984) R_1-responses of the trigemino-facial reflex in lesions extrinsic to the brain stem. Ital J Neurol Sci 5: 41–44

124. Rossi B, Buonaguidi R *et al* (1979) Blink reflexes in posterior fossa lesions. J Neurol Neurosurg Psychiat 42: 465–469

125. Rowe MJ, Carlson C (1980) Brain stem auditory evoked potentials in postconcussion dizziness. Arch Neurol 37: 679–683

126. Rumpl E, Gerstenbrand F et al (1982) Some observations on the blink reflex in posttraumatic coma. EEG Clin Neurophysiol 54: 406–417

127. Salomon G, Elberling C, Tos M (1979) Combined use of electrocochleography of acoustic neuromas. Rev Laryng 100: 697–707

128. Sancesario G, Pozzessere G et al (1984) Prognostic evaluation of brain stem haematomas: the role of CT scan and brain stem auditory evoked potentials. Acta Neurol Scand 70: 396–406

129. Scherg M, von Cramon D (1986) Evoked dipole source potentials of the human auditory cortex. EEG Clin Neurophysiol 65: 344–360

130. Scherg M, von Cramon D (1985) A new interpretation of the generators of BAEP waves I–V. EEG Clin Neurophysiol 62: 290–299

131. Scherg M (1987) Dipolar source of the early auditory evoked potentials. Clin Neurol Neurosurg [Suppl 1] 89: 18

132. Seales DM, Rossiter VS, Weinstein ME (1979) Brain stem auditory evoked responses in patients comatose as a result of blunt head trauma. J Trauma 19: 347–352

133. Selters WA, Brackmann DE (1977) Acoustic tumour detection with brain stem electric response audiometry. Arch Otolaryngol 103: 181–187

134. Selters WA, Brackmann DE (1979) Brain stem electric response audiometry in acoustic tumour detection. In: House WF, Luetje CM (eds) Acoustic tumours, vol 1, Diagnosis. University Park Press, Baltimore, pp 225–235

135. Serrats AF, Parker SA, Merino-Canas A (1976) The blink reflex in coma and after recovery from coma. Acta Neurochir (Wien) 34: 79–97

136. Shakhnovich AR, Thomas JG et al (1979) A study of the mechanisms of comatose patients. Seara M Neurocirurg 3: 269–300

137. Shanon E, Gold S, Himmelfarb MZ (1981) Auditory brain stem responses in cerebellopontine angle tumours. The Laryngoscope 91: 254–259

138. Sohmer H, Feinmesser M (1967) Cochlear action potentials recorded from the external ear in man. Ann Otol 76: 427–435

139. Sohmer H, Feinmesser M, Szabo G (1974) Sources of electrocochleographic responses as studied in patients with brain damage. EEG Clin Neurophysiol 37: 663–669

140. Sontheimer D, Caird D, Klinke R (1985) Intra- and extracranially effects of interaural time and intensity difference. EEG Clin Neurophysiol 61: 539–547

141. Starr A (1976) Auditory brain stem responses in brain death. Brain 99: 543–554

142. Starr A, Achor LJ (1975) Auditory brain stem responses in neurological disease. Arch Neurol 32: 761–768

143. Starr A, Hamilton AE (1976) Correlation between confirmed sites of neurological regions and abnormalities of far-field auditory brain stem responses. EEG Clin Neurophysiol 41: 595–608

144. Stephens SDG, Thornton ARD (1976) Subjective and electrophysiological tests in brain stem lesions. Arch Otolaryngol 102: 608–613

145. Stern BJ, Krumholz A et al (1982) Evaluation of brain stem stroke using brain stem auditory evoked responses. Stroke 13: 705–711

146. Stockard JJ, Rossiter VS (1977) Clinical and pathological correlates of brain stem auditory response abnormalities. Neurology 27: 316–325

147. Stockard JJ, Stockard JE, Sharbrough FW (1978) Nonpathologic factors influencing brain stem auditory evoked potentials. Am J EEG Technol 18: 177–209

148. Stockard JJ, Sharbrough FW (1979) Brain stem auditory response patterns associated with cerebellopontine angle lesions. EEG Clin Neurophysiol 46: 15–19

149. Stöhr M, Dichgans D et al (1982) Evozierte Potentiale. Springer, Berlin Heidelberg New York

150. Straschill M (1980) Orbicularis-oculi-Reflex mit fehlender früher Komponente und normaler später Reaktion bei einem Patienten mit intrapontinem Tumor. Z EEG-EMG 11: 19–20

151. Struppler A, Dobbelstein H (1963) Elektromyographische Untersuchung des Glabellareflexes bei verschiedenen neurologischen Störungen. Nervenarzt 34: 347–353

152. Sutton LN, Frewen T et al (1982) The effects of deep barbiturate coma on multimodality evoked potentials. J Neurosurg 57: 178–185

153. Sutton LN, Cho BK et al (1986) Effects of hydrocephalus and increased intracranial pressure on auditory and somatosensory evoked responses. Neurosurg 18: 756–761

154. Szabo M, Krajcso T (1987) Brain stem auditory evoked potential findings in patients with vascular brain stem lesions. Clin Neurol Neurosurg [Suppl 1] 82: 71

155. Tackmann W, Ettlin T (1982) Electrically, acoustically and visually elicited blink reflexes. II. Their relation to visual evoked potentials and auditory brain stem evoked potentials in the diagnosis of multiple sclerosis. Eur Neurol 21: 264–269

156. Takada M, Itoh K et al (1984) Distribution of premotor neurons for orbicularis oculi motoneurons in the cat, with particular reference to possible pathways for blink reflex. Neurosci letters 50: 251–255

157. Takeuchi Y, Nakano K et al (1979) Mesencephalic and pontine afferent fiber system to the facial nucleus in the cat: a study using the horseradish peroxidase and silver impregnation techniques. Exp Neurol 66: 330–343

158. Tebecis AK (1970) Effects of monoamines and amino acids on medial geniculate neurones of the cat. Neuropharmacology 9: 381–390

159. Tebecis AK (1967) Are 5-hydroxytryptamine and noradrenaline inhibitory transmitters in the medial geniculate nucleus? Brain Res 6: 780–782

160. Terkildsen K, Huis in't Veld F, Osterhammel P (1977) Auditory brain stem responses in the diagnosis of cerebellopontine angle tumours. Scand Audiol 6: 43–47

161. Tokunaga A, Oka M et al (1958) An experimental study on facial reflex by evoked electromyography. Med J Osaka Univ 9: 397–411

162. Tsubokawa T, Nishimoto H et al (1980) Assessment of brain stem damage by the auditory brain stem response in acute severe head injury. J Neurol Neurosurg Psychiat 43: 1005–1011

163. Tsutsui T, Avila A et al (1986) The effects of a supratentorial mass lesion on brain stem auditory evoked potentials and short latency somatosensory evoked potentials. Neurol Res 8: 13–17

164. Uziel A, Benezech J (1978) Auditory brain stem responses in comatose patients: relationship with brain stem reflexes and levels of coma. EEG Clin Neurophysiol 45: 515–524

165. Uziel A, Benezech J et al (1982) Clinical applications of brain stem auditory evoked potentials in comatose patients. In: Courjon J et al (eds) Clinical applications of evoked potentials in neurology. Raven Press, New York, pp 195–202

166. Velasco M, Velasco F (1987) Sources of some vertex auditory

evoked potentials: intracranial study in man. Clin Neurol Neurosurg [Suppl 1] 89: 76

167. Voordecker P, Zegers de Beyl D, Brunko E (1987) Selective unilateral absence or attenuation of wave V of BAEP with intrinsic brain stem lesions. Clin Neurol Neurosurg [Suppl 1] 89: 114

168. Wielaard R, Kemp B (1979) Auditory brain stem evoked responses in brain stem compression due to posterior fossa tumours. Clin Neurol Neurosurg 81: 185–193

169. Zappia M, Lüders H (1987) BAEP's recorded from the basal surface of the temporal lobe: laterization of the components. Clin Neurol Neurosurg [Suppl 1] 89: 132

170. Zierski J, Kurzaj E et al (1983) Cerebral blood flow in the brain stem during increased ICP. In: Intracranial pressure V. Springer, Berlin Heidelberg New York

Authors' address: Professor Dr. Norfrid Klug und PD Dr. György Csécsei, Department of Neurosurgery, University of Giessen, Klinikstrasse 29, D-6300 Giessen, Federal Republic of Germany.

Acta Neurochirurgica, Suppl. 40, 95–116 (1987)

Blood Flow in Brain Structures During Increased ICP

Jan Zierski

Department of Neurosurgery, University of Giessen, Federal Republic of Germany*

Contents

Summary

The effect of a supratentorial expanding mass lesion and of uniform increase of ICP on regional cerebral blood flow was examined in 31 cats. The blood flow was measured using the radioactive microsphere technique and continuous ICP increase was produced by inflating an extradural balloon or by infusion of mock CSF into subarachnoid lumbar space. Four additional animals in whom no ICP rise was produced were used as controls; several blood flow measurements were performed at different ICP levels and after sudden ICP release. The analysis of the data obtained revealed that intracranial hyperten-

* Present address: Department of Neurosurgery, Hospital Neu-koelln, Berlin West.

sion caused inhomogenous pattern of blood flow change with compartmentalization of flow between supra- and infratentorial structures connected with cisternal herniation. The flow decrease may correspond to the craniocaudal pressure gradients in the brain stem. Irrespective of the method used to produce intracranial hypertension the blood flow in the lower brain stem was less susceptible to diminished perfusion pressure. Sparing of cerebral blood flow in the lower brain stem during progressive brain compression can be explained by compartmentalization. The ranking of regions at cerebral perfusion pressure below 60 mm Hg was similar for the lower brain stem regions independently of the method which was used to increase the ICP. This suggests that when CBF becomes reduced due to increase of ICP the perfusion favours the areas where neurons related to control of circulation are located. Diffuse increase of ICP produced no interhemispheric differences in the blood flow. These differences were detected when balloon compression was used. Asymmetry of perfusion in the brain stem structures was not observed.

During continuously increasing ICP an increase of blood pressure taking place before pupillary dilatation occurred was not caused by medullary ischaemia. If the pressure continued to increase the vasopressor response occurring after pupillary dilatation took place did not improve the cerebral blood flow. Increase of cerebral perfusion followed a sudden release of ICP. In an experimental animal subjected to unilateral compressive lesion producing tentorial herniation, hyperperfusion involved especially the thalamus and the midbrain with relative flow decrease in the lower brain stem.

Keywords: Intracranial pressure; cerebral blood flow; brain herniation; brain stem blood flow.

Introduction

The effects of raised intracranial pressure (ICP) upon cerebral blood flow (CBF) and focal flow in selected areas have been extensively studied in the past with the aim of understanding the regulation of CBF. When ICP increase was produced by stepwise infusion of fluid into the subarachnoid space the CBF was found to be maintained despite raising ICP and decreasing cerebral perfusion pressure (CPP) expressed as difference between ICP and mean systemic arterial blood pressure (MABP)[23, 32, 91]. Passive decrease of CBF occurred when CPP was reduced to the values of approximately 40–50 mm Hg.

When balloon inflation was used to increase ICP, the flow was found to be reduced as soon as the ICP rose[33, 46] but this result was not obtained by Johnston et al.[34]. With balloon expansion over cerebral hemispheres in primates Johnston et al.[33] found no significant differences between the hemispheric flows and no substantial interhemispheric pressure gradients. Symon et al.[71] using more rapid expansion of extradural balloon found that pressure gradients developed between the hemispheres and that the regulatory capacity of blood flow in the hemisphere subjected to greater compression is more exhausted than on the opposite side. This leads to establishment of differential flow pattern for a short time during the critical phase of compression. The pressure gradients between the supra- and infratentorial compartment occurring with progression of cisternal block[4, 42] are bound to produce intercompartmental CPP differences and influence the blood flow.

In the majority of studies relating increased ICP to CBF in animal experiments the total CBF or flow in a few selected areas has been measured[23, 42, 58, 71]. There is little information about measurement of blood flow in the brain stem due to increased ICP and about changes in regional distribution of blood flow which may occur in the phase of rapidly increasing ICP[2, 70]. The aim of the present study was to investigate the effect of intracranial pressure increase upon regional cerebral blood flow with special attention devoted to the flow in deep brain structures.

Material and Methods

35 adult cats of both sexes weighting 2.4–3 kg were used. Anaesthesia was induced with intraperitoneal pentobarbital sodium in a dosage of 32 mg/kg body weight. The animals were tracheotomized and paralysed with tubocurarine 0.1–0.3 mg/kg in divided doses. Ventilation with room air supplemented by oxygen was controlled using a Starling-pump, tidal volume was adjusted according to end-tidal CO_2 and arterial blood gas estimations. ABP was measured from a catheter introduced into the abdominal aorta via the left femoral artery. Central venous pressure (CVP) was measured from a catheter inserted into the inferior vena cava via the femoral vein (Brown Comp. No. 4 int. diam. 1 mm, ext. diam. 1.5 mm). Left sided thoracotomy was performed, a polyethylene catheter of the above size was introduced into the left atrium, secured with sutures and the chest was tightly closed. This catheter was used for injection of microspheres for CBF estimations. The right femoral artery was

catheterized (Brown Comp. No. 3. int. diam. 0.8 mm, ext. diam. 1.3 mm), the catheter was advanced to the abdominal aorta and attached to the automatic pump (Unita-2, Braun, Melsungen) for withdrawal of reference blood samples during CBF measurements. Arterial blood gases were examined at frequent intervals and every time when CBF measurements were taken. Samples were analysed in Cornig Blood Gas Analyser. Hemoglobin was estimated at the same intervals. The animals were placed in a stereotactic frame on a heated pad and temperature was monitored continuously through s subcutaneous probe (Hartman and Brown Comp.).

Electrocardiogram was recorded from needle electrodes placed at extremities. Electroencephalogram was monitored from frontal and parietal skull leads. Intracranial pressure was measured from the extradural space on the left side by means of calibrated transducer Statham SP-50 mounted in a special screw.

All pressure recordings were referred to the interaural line. The pressure data was displayed on a 12 channel variable speed chart recorder (Mingograph, Siemens) at paper speed 1.5 mm/sec changing to 30 mm/sec during the period of isotope injection. Pressure values and heart rate were displayed additionally on a trend recorder (12 channel Kompensograph, Siemens).

1. Blood Flow Measurement

The blood flow was measured by injection of radioactive microspheres 15 ± 3 μ in diameter (Nen-Track, New Englang Nuclear) according to the method developed by Rudolph and Heymann[62] and using the reference sample method[48]. According to the principles of measurement blood flow with microspheres established by Buckberg et al.[8] 1×10^6 microspheres were injected into the left atrium within 20–30 sec. Blood flow for the reference count was withdrawn at constant rate of 2 ml/min for 2 minutes starting immediately before the injection. Repeated CBF measurements (4–5) were performed using the following isotopes: 141-Ce, 113-Sn, 103-Ru, 95-Nb, 46-Sc, 51-Cr, 153-Gd. The microspheres suspended in isotonic saline solution were continuously shaken prior to injection and examined using a light microscope to ensure that the spheres were dispersed.

Nuclide counting was performed with a 256 pulse-height-analyser using an intrinsic Ge-well type detector (Princeton Gamma Tech. Inc.) mounted in the lead shield of a modified automatic sample changer. Counting procedure was done in the Max-Planck-Institute of Cardiac Research in Bad Nauheim. The computer program developed in this institute[65, 75, 76] was used for the separation of isotopes, correction of Compton-Continuum, control of the sample changer, data accumulation and calculations. The data was processed with an interactive computer programme running on a PDP 11/45. The activities were read in the computer together with information about the isotopes and the corresponding calibration flow values and were converted into the corresponding values in blood flow (ml/min 100 g)[75].

2. Experimental Protocol

After establishing all measurement lines and stabilization of the animal for one hour control CBF measurement was performed. ICP was then raised in 15 animals by the continuous inflation of the latex balloon at the rate of 0.05–0.1 ml/min. The balloon was placed through a small craniectomy in the right fronto-temporal region and the opening sealed with acrylic. The inflated volume was 1.7–3.5 ml producing a rise of ICP to values around 80–100 mm Hg within 40–50 min.

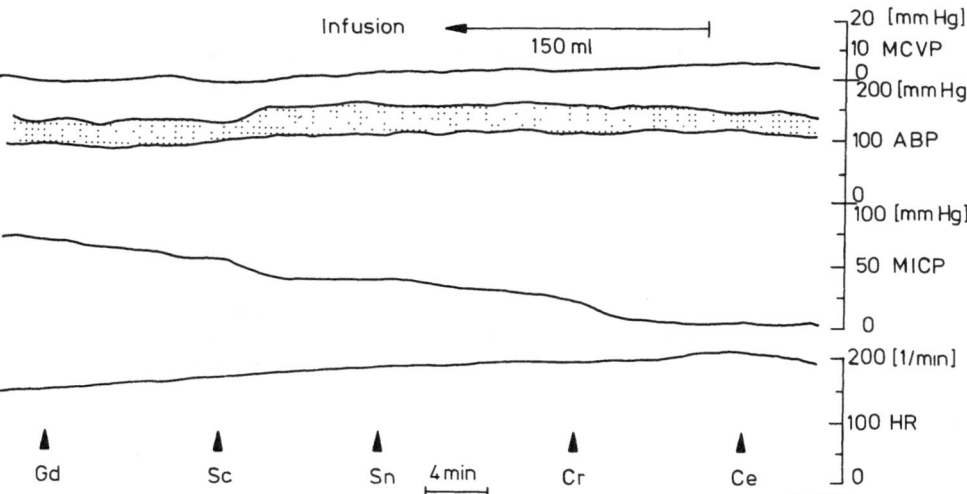

Fig. 1. Example of the course of ICP increase and timing of CBF measurements in a single animal. ICP raised to 75 mmHg in 52 min

Results

1. Control Values

Four control cats were operated upon. At random intervals four CBF measurements were performed on each of them without producing ICP increase to check the influence or preparation, multiple microsphere injections, blood loss and duration of anaesthesia upon CBF. The intraindividual standard deviation for total CBF was around 3.0 ml/min/100 g, the maximum range was 8 ml/min/100 g. When regional CBF samples were compared—in a few instances larger ranges were found, particularly in areas with a high flow. The difference

In 16 animals pressure increase was produced by infusion of mock-CSF through a lumbar catheter advanced to the midthoracic region. Variable flow infusion pump (Perfursor Braun, Melsungen) was used in order to obtain similar pressure time curves as in the balloon group (Fig. 1). The mean infusion rate was 3.1 ± 1.3 (SD) ml/min for the whole period of the experiment but varied according to needs during the time.

Serial CBF measurements were performed at selected periods of constant pressure levels separated by 10–20 mm Hg. If during the injection of microspheres or withdrawal period for blood sampling mean ICP changed by more than 3 mm Hg, the measurement was rejected.

The animals were sacrificed by intracardiac injection of KCl. The brain and spine were removed, inspected macroscopically, fixed in 4% formaline for 2–3 days and dissected. The brain was divided into 50–52 blocks weighing between 150 and 900 mg. The only exception was a sample of choroid plexus of the lateral ventricle which weighed on the average 70–80 mg. Spine, kidneys and heart were also taken for estimation of flow and cut into blocks. Blood samples were divided in aliquots. Total and regional blood flow were expressed as weighted means of corresponding tissue blocks.

In some animals blood flow measurements were performed after releasing ICP by rapid deflation of balloon or withdrawal of CSF.

between flow in both hemispheres amounted to maximum 8 ml/min/100 g in two cats and was less than 3 ml/min/100 g in the remaining two animals (Table 1). No differences in the flow between the right and left side were disclosed at 99% confidence interval in the midbrain, upper and lower medulla. For the thalamus the deviation of slope from the 99% confidence interval was minimal (Table 2).

No systemic reactions or EEG changes were observed which could be connected with the injection of 1×10^6 microspheres. The values for regional and total CBF obtained from control operated cats and from the first measurement of the remaining animals after they had achieved the steady state are shown in Table 3.

The mean CBF of the total brain was 37.8 ± 2.3 (M ± S.E.M.) ml/min/100 g. Most of the brain regions examined had a mean CBF in a narrow range from 37–45 ml/min/100 g. The largest flow was present in the choroid plexus with a considerable variation probably caused by small weight (< 100 mg) of the sample. The caudate nucleus also showed a relatively high flow in comparison with other areas. As expected the lowest values were recorded in samples of pure white matter such as corpus callosum and optic chiasm. The mean values for the sample of cortex were also lower than expected but the samples contained a rather large proportion of subcortical white matter.

2. Effect of ICP upon Systemic Measurements

Table 4 shows the values of systemic measurements for the balloon group and infusion group at the steady state compared to the values obtained when measurements were taken at CPP below 60 mm Hg.

On the average there was a decrease in MABP and

Table 1. *Total brain, regional blood flows and systemic parameters in four control operated cats with 4 estimations of flow during the steady state.* Microspheres injected at random over 30–60-minute period

Number of estimations	C 1	C 2	C 3	C 4
	4	4	4	4
BT	23.0 ± 3.4*	28.2 ± 2.6	42.2 ± 3.2	44.5 ± 3.1
HL	22.0 ± 2.1	30.0 ± 3.4	41.2 ± 1.2	40.0 ± 4.4
HR	20.5 ± 2.3	29.0 ± 2.4	42.2 ± 4.1	46.5 ± 3.5
Frontal	25.2 ± 3.5	37.0 ± 2.9	51.0 ± 2.7	46.7 ± 7.2
Temporal	16.2 ± 4.9	15.7 ± 0.9	37.0 ± 4.8	42.7 ± 2.0
Occipital	25.5 ± 1.2	37.5 ± 3.0	46.7 ± 1.5	46.2 ± 3.4
Thalamus	32.0 ± 10.3	30.0 ± 2.4	39.7 ± 4.9	46.5 ± 3.4
Hypothalamus	23.2 ± 10.9	15.5 ± 1.7	51.7 ± 11.8	45.5 ± 3.0
N. caudatus	68.0 ± 12.1	49.7 ± 5.2	55.7 ± 3.5	103.5 ± 16.0
C. Call.	14 ± 8.6	10.5 ± 1.2	23.7 ± 7.1	24.5 ± 2.5
Midbrain	30.2 ± 8.0	26.7 ± 2.5	45.2 ± 2.2	46.2 ± 2.5
Pons	26.2 ± 7.0	15.5 ± 1.2	43.7 ± 3.9	47.0 ± 1.4
Medulla	29.0 ± 7.6	20.7 ± 1.5	47.7 ± 4.5	46.2 ± 2.2
Cerebellum	28.7 ± 5.5	27.2 ± 4.7	44.0 ± 2.4	63.5 ± 9.9
Plexus chor.	148.0 ± 37.3	103.5 ± 13.9	262 ± 26.7	255.5 ± 25.3
Spinal cord				
cervical	9.5 ± 2.0	16.5 ± 1.9	14.2 ± 0.9	15.7 ± 0.7
thoracic	8.5 ± 2.6	13.0 ± 2.9	13.2 ± 2.2	13.0 ± 1.6
lumbal	12.7 ± 2.0	13.7 ± 2.2	10.0 ± 0.8	23.0 ± 0.8
Heart	173.7 ± 17.0	329.2 ± 62.0	353.5 ± 53.2	278.0 ± 37.2
Kidney left	204.5 ± 24.5	137.5 ± 41.6	150.2 ± 12-9	140.0 ± 19.2
Kidney right	200.0 ± 36.8	135.5 ± 31.1	133.5 ± 40.1	131.0 ± 22.4
ABP	110 ± 8	145 ± 5	166 ± 4	145 ± 7
HR	140 ± 1	175 ± 5	222 ± 13	234 ± 11
CVP	4.4 ± 1.2	6.8 ± 0.8	6.6 ± 1.6	1.4 ± 0.8
pCO_2	30.4 ± 2.2	34.4 ± 1.8	28.3 ± 1.1	32.2 ± 2.6
pH	7.31 ± 0.07	7.32 ± 0.02	7.32 ± 0.01	7.33 ± 0.02

* M ± S.D.

Table 2. *Comparison of CBF of the right and left side in deep brain structures.* Control values (45 measurements in 35 cats)

Region	Mean rCBF ml/min/100 g left		Correlation coefficient R.	Slope	Intercept
Thalamus	41.5	40.5	0.92	0.83 ± 0.13	6.05 ± 5.95
Midbrain	42.3	41.4	0.91	1.03 ± 0.18	−2.36 ± 8.14
Upper Med. obl.	37.8	39.9	0.82	0.88 ± 0.23	6.66 ± 9.45
Lower Med. obl.	25.6	26.6	0.86	0.95 ± 0.21	2.33 ± 6.00

Table 3. *CBF control values (ml/min/100 g).* N = 45; 35 animals

Region	Mean	S.E.M.
Chor. plexus	249.2 ± 60.3	
N. caudatus	70.7 ± 4.2	
Frontal	44.8 ± 3.2	
Occipital	42.1 ± 3.0	
Midbrain	41.8 ± 2.8	
Thalamus	40.9 ± 3.0	
Cerebellum	39.9 ± 2.3	
Medulla obl. upper	38.8 ± 2.6	
Hypothalamus	37.8 ± 3.4	
Temporal	37.3 ± 2.5	
Pons	33.7 ± 2.4	
Grey matter samples	28.0 ± 2.1	
Medulla obl. lower	28.0 ± 2.0	
Corp. call. + chiasm	20.5 ± 1.3	
Brain total	37.8 ± 2.3	
Hemisphere left	36.8 ± 2.3	
Hemisphere right	37.5 ± 2.4	

of the pupil ipsilateral to the balloon occurred at ICP levels between 52 and 86 mm Hg (mean ICP 77 ± 4.7 S.E.M. mm Hg) and EEG became flat or isoelectric at an average CPP of 12 ± 2.5 S.E.M. mm Hg; the highest value of CPP at which EEG became flat was 25 mm Hg in 2 animals. In all the remaining cases CPP dropped below 18 mm Hg at the moment of flat EEG.

In the infusion group pupillary dilatation occurred in only 3 animals at ICP levels of 84, 96 and 100 mm Hg respectively. The EEG became flat in only 2 animals at CPP levels of 35 and 55 mm Hg respectively.

3. Macroscopic Changes

Tentorial herniation and midline displacement was present in all animals in the balloon group when the brain was removed. When brains were sectioned it was found that in the balloon group 1 animal had a haemorrhage in the midbrain and 4 other subarachnoid, cortical and subcortical haemorrhages at the site of the balloon placement. Three of these animals belonged to the group, which was allowed to survive for 20–30 min after deflation of the balloon.

In animals of the infusion group in one case haemorrhage was disclosed in the caudate nucleus and in one over the left posterior hemisphere. In 2 further cats spinal subarachnoid haemorrhage at the lumbo-thoracic level were found.

HR in the balloon group and a MABP increase in the infusion group (Fig. 2).

There was no marked change in other parameters. In all the animals in the balloon group maximal dilatation

Table 4. *Systemic measurements (mean ± S.E.M.)*

	Balloon		Infusion	
	Steady state (N = 15)	Final stage (Cpp < 60 mm Hg) (N = 13)	Steady state (N = 16)	Final stage (CPP < 60 mm Hg) (N = 8)
MABP (mm Hg)	137 ± 6	121 ± 12	139 ± 4	156 ± 7
HR (1/min)	175 ± 14	148 ± 10	191 ± 8	170 ± 11
pO_2 (mm Hg)	105 ± 7	127 ± 8	120 ± 6	139 ± 10
pCO_2 (mm Hg)	32.4 ± 0.9	30.8 ± 1.2	29.9 ± 0.7	29.7 ± 1.0
pH	7.30 ± 0.02	7.34 ± 0.02	7.32 ± 0.01	7.29 ± 0.03
Endresp. CO_2 (%)	3.6 ± 0.2	3.1 ± 0.2	3.9 ± 0.3	3.6 ± 0.3

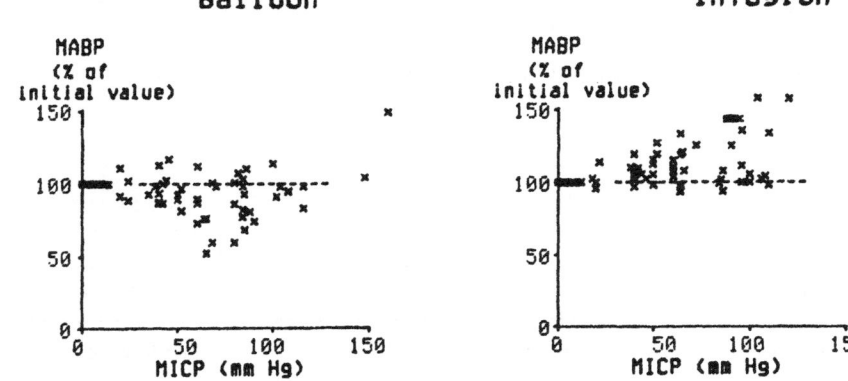

Fig. 2. Mean arterial blood pressure change during the experiment expressed as per cent of initial value. Note the tendency to ABP decrease in the balloon group and ABP increase in the infusion group

4. rCBF in the Initial Phase of ICP Increase

In order to evaluate the rCBF in the initial phase of ICP increase, flow values in 20 regions recorded before pressure increase were compared with values measured when ICP reached 40 mm Hg. Nine animals in the balloon group and 13 animals in the infusion group were available for this comparison. The flow was considered unchanged, if the values did not differ by more than 15%, otherwise it was classified as flow increase or flow decrease respectively.

In the majority of animals there was no uniform pattern of rCBF change. Decrease of CBF in all regions examined as found in 4 out of 22 animals and flow decrease in more than 50% of regions in a further 9. In 9 out of 22 cats the flow remained unchanged or increased in more than 50% regions during the initial phase of ICP increase.

The number of animals was however too small to detect significant differences in the behaviour of regional CBF between the balloon and infusion group.

In the s.c. autoregulatory range of CPP change at values of CPP above 60 mm Hg there was a significant correlation (p < 0.01) between total CBF and CPP in both the balloon group and the infusion group (r = 0.59/r = 048) (Fig. 3). At the same range of CPP there was a significant negative correlation (p < 0.01, r = —0.52) between CBF and ICP change in the balloon group.

No such relation was found in the infusion group (r = —0.14, n.s.).

5. Effect of ICP upon Regional CBF

5.1. Comparison of Regional Flows with Total Brain Flow

In order to compare interregional blood flow changes within each group and between both groups, regional flow values were plotted against the respective total brain flow and expressed as a percentage of the respective control value. The distribution of points which represented the relative flows at the same measurement in the regions compared, were then analysed in respect to the identity line (y = x) at significance levels p = 0.01. This provided information

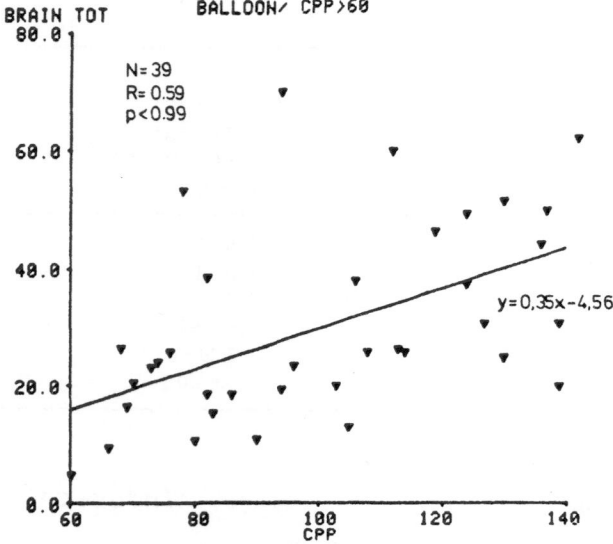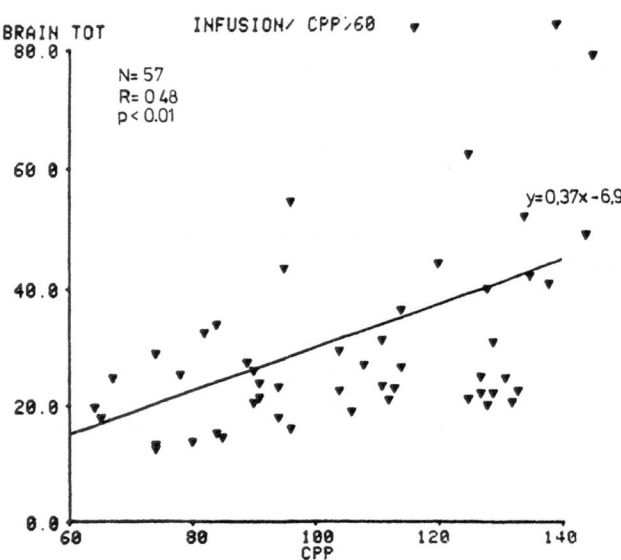

Fig. 3. Relation between CBF and CPP at CPP above 60 mmHg

Table 5 a. *Changes of regional blood flow in relation to total CBF.* Balloon group

Not different	More rapid*	Less rapid*
N. caudatus	ipsilateral	contralateral
Thalamus	hemisphere	hemisphere
Hypothalamus		midbrain
C. callosum		pons
+ chiasma		upper medulla
		lower medulla
		cerebellum

* Significant in comparison with total CBF p < 0.01.

Table 5 b. *Changes of regional blood flow in relation to total CBF.* Infusion group

Not different	More rapid*	Less rapid
Left hemisphere	none	midbrain
Right hemisphere		cerebellum
N. caudatus		
Thalamus		
Hypothalamus		
C. callosum + chiasma		
Upper medulla		
Lower medulla		

* Significant in comparison with total CBF p < 0.01.

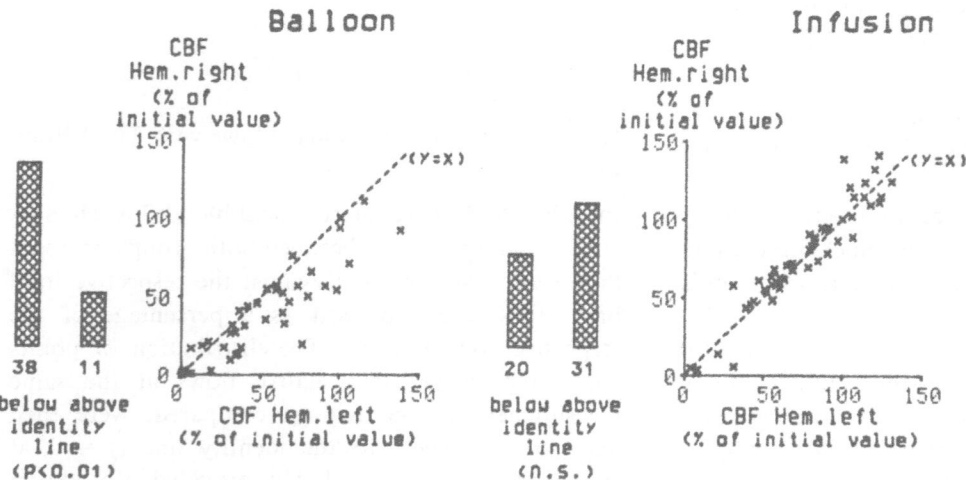

Fig. 4. Comparison between the change of blood flow in the hemispheres in the balloon and infusion group. The flow diminishes more on the compressed side. No such difference in the infusion group. Note that in several instances in the infusion group the blood flow increases above the control level when ICP is raised. This is exceptional in the balloon group

as to the rate and degree of flow change in the given region in comparison with the total cerebral flow. Cumulative results are shown in Table 5. A more uniform pattern of flow change is seen in the infusion group than in the balloon group. With progressing inflation of the balloon the flow decreased more rapidly in the ipsilateral hemisphere and less rapid in the contralateral hemisphere and brain stem (Fig. 4).

Redistribution of regional flow in the balloon group became evident at ICP levels of 40–60 mm Hg. The decrease of flow in the ipsilateral hemisphere was matched by a relative flow increase in the pons and medulla by appr. 40–70%.

5.2. Comparison of Left and Right Side Flows

The same method of comparison (distribution along identity line) was used. Cumulative data for both groups is shown in Table 6.

In the balloon group significant differences of CBF reduction were seen for the temporal and occipital regions, nucleus caudatus and the hemispheres. For samples of the frontal lobe the same trend with greater impairment of flow on the side of the balloon was found, but was not significant at the 0.01 level. There were no differences in the blood flow of the brain stem structures in the samples of the right and left side (Fig. 5).

In contrast to the balloon group the regional decrease of CBF in the infusion group was almost uniform for all the regions studied. In some animals the flow in the brain stem was preserved or increased at the time when the hemispheric flow dropped (Fig. 6).

In both the balloon and the infusion group individual cats showed regional variability of CBF with the

Table 6. *Changes of regional blood flow in relation to the opposite site during increasing ICP*

Balloon group		Blood flow change	
regions compared		Not different	Different*
Hemisphere	R/L		×
Frontal	R/L	×	
Temporal	R/L		×
Occipital	R/L		×
N. caudatus	R/L		×
Thalamus	R/L	×	
Midbrain	R/L	×	
Upper medulla	R/L	×	
Lower medulla	R/L	×	
Cerebellum	R/L	×	

* Significant in comparison with the opposite side $p < 0.01$.

flow increasing in some regions and decreasing in others.

5.3. Brain Stem Flow

Change of flow during the experiment was compared for caudate nucleus, thalamus, midbrain, pons, upper and lower medulla oblongata and cerebellum. In the balloon group significant interregional differences ($p < 0.01$) in the degree of CBF reduction, successively in rostro-caudal direction, were found with the exception of the lower brain stem. Within the posterior fossa structures there was also no significant difference in the degree of flow reduction between the cerebellum and pons but there was a difference between cerebellum and upper medulla. The same comparison applied to the infusion group showed no significant interregional

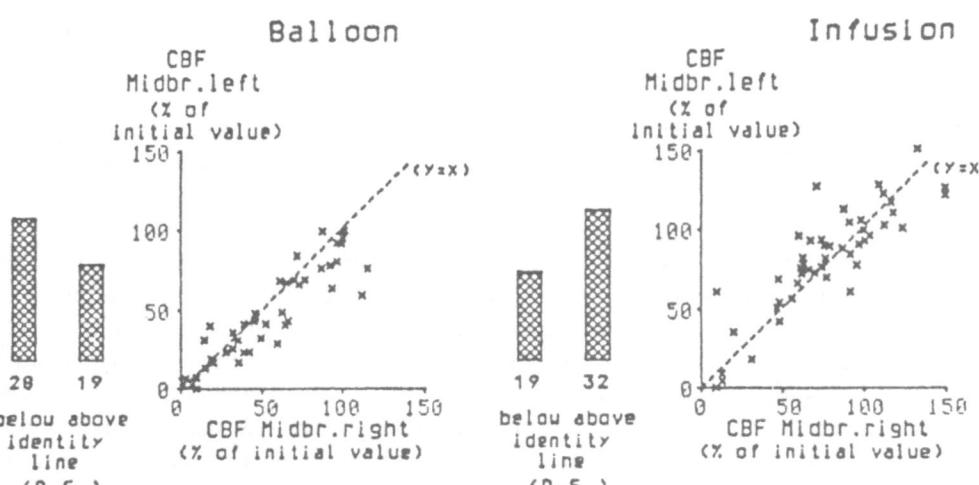

Fig. 5. Comparison between the change of flow in the midbrain for the balloon and infusion group. No differences are disclosed

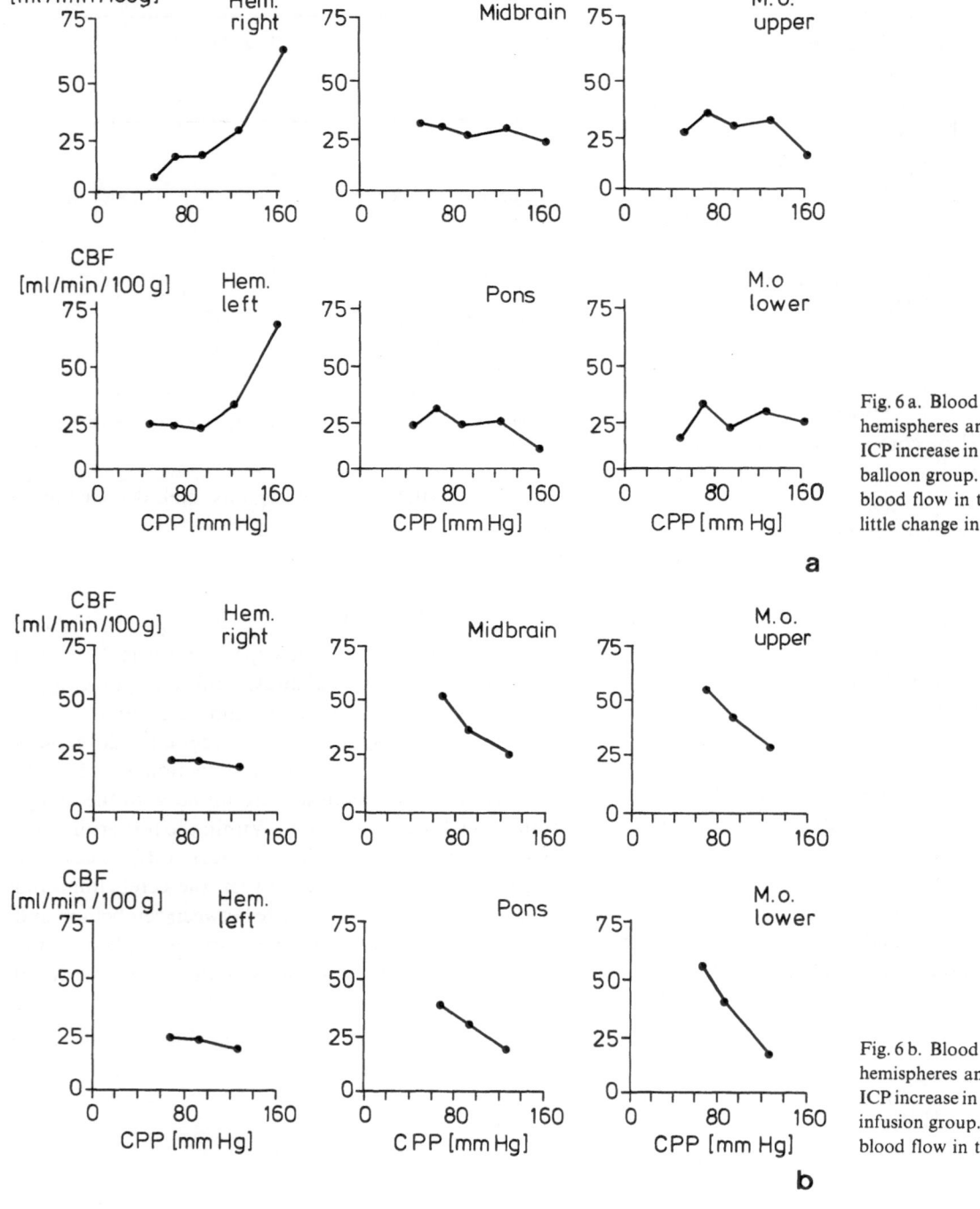

Fig. 6 a. Blood flow in the cerebral hemispheres and brain stem during ICP increase in a single animal of the balloon group. Note the decrease of blood flow in the hemispheres with little change in the brain stem

Fig. 6 b. Blood flow in the cerebral hemispheres and brain stem during ICP increase in a single animal of the infusion group. Note the increase of blood flow in the brain stem

differences with a single exception of flow in the cerebellum versus upper medulla. The flow in the cerebellum was less affected (Fig. 7).

In both the groups, however, the change of CBF was smaller in the brain stem than in the supratentorial structures. This is shown in Fig. 8 where the final CBF (at CPP below 60 mm Hg) is expressed in per cent of initial value and ranked accordingly.

In the balloon group, at CPP < 60 mm Hg the flow in the medulla, cerebellum and pons was reduced to approximately 50 per cent of initial value, in the midbrain to 27 per cent and in the majority of other regions to below 12 per cent of the initial value. In the infusion group CBF for all the regions studied was reduced on average to 40–65 per cent of the initial value.

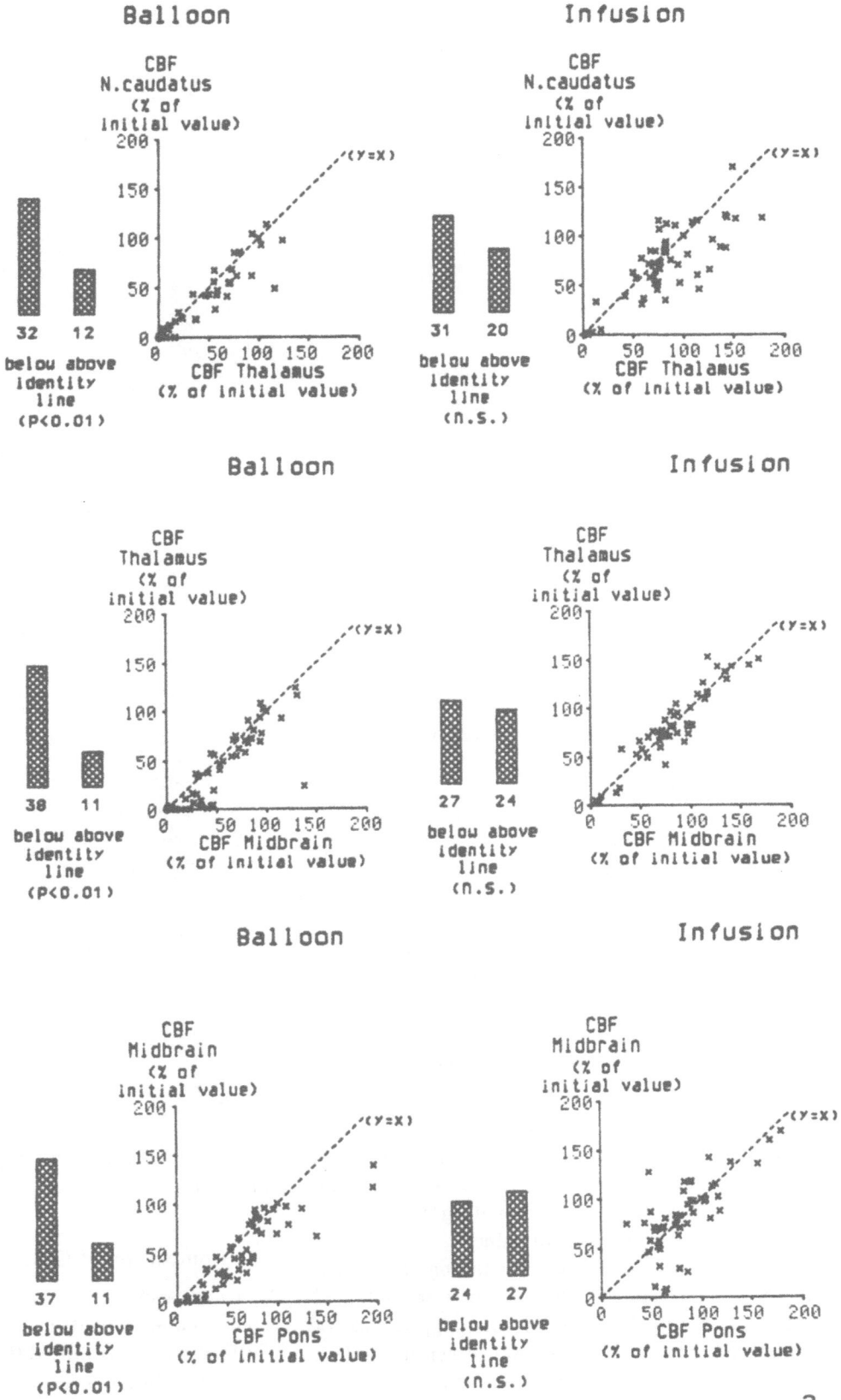

Fig. 7a. Interregional comparison of flows in central brain structures

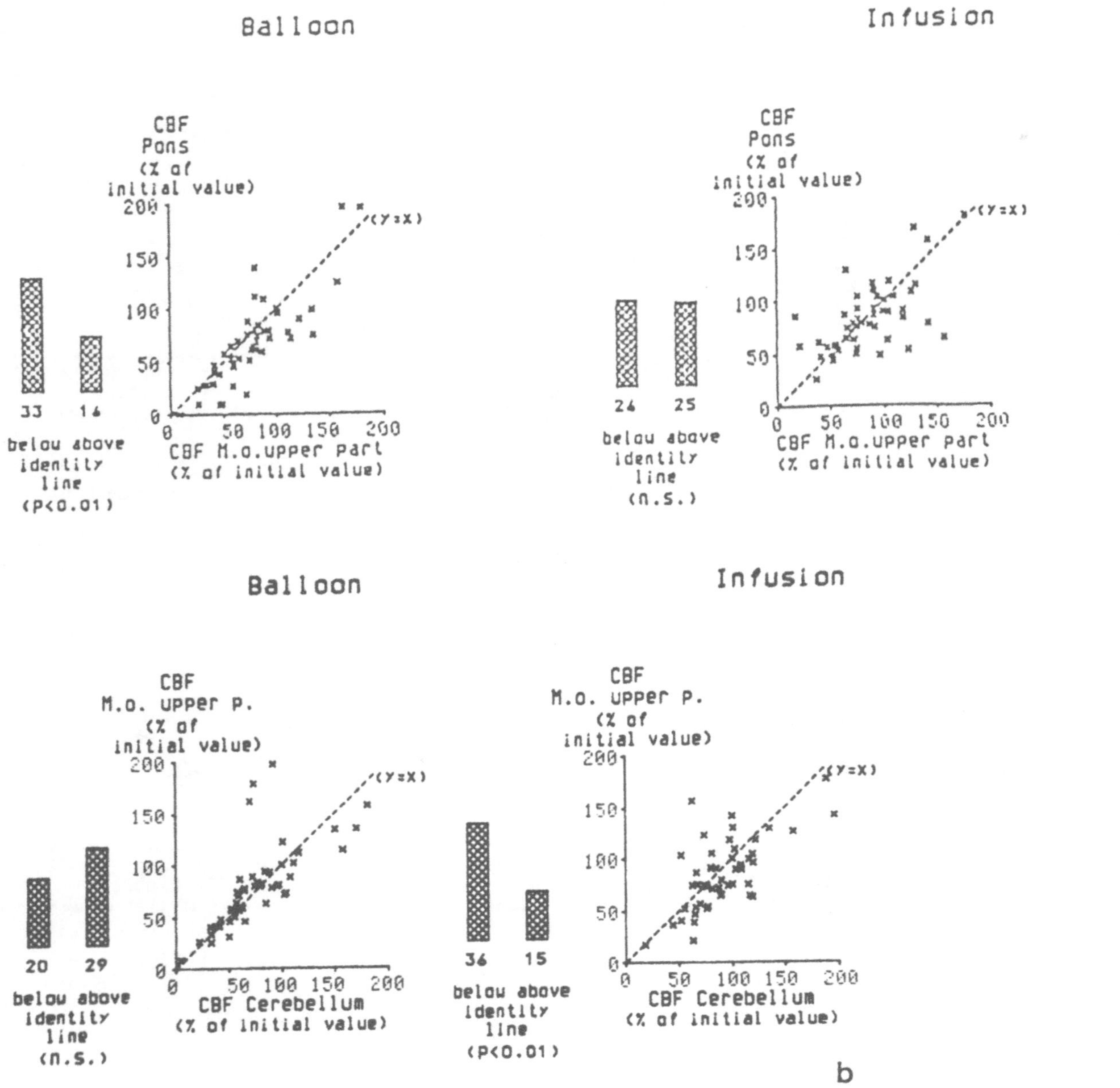

Fig. 7 b. Interregional comparison of flows in central brain structures

Flow values in the midbrain after right sided anisocoria developed were obtained in 13 animals. The midbrain flow at that time was 12.0 ± 10 mm/100 g/min equal to 26 ± 18 per cent of the control value.

In only 7 instances, however, the timing of two flow estimations was such that flow values were obtained immediately before and shortly after appearance of pupillary dilatation. The values are shown in Table 7. The mean time between flow measurements was 8 ± 2 min (range 5–11 min). It is evident that the development of anisocoria is associated with rapid,

profound decrease of already compromised flow in the midbrain.

5.4. Comparison of Control Flow with Final Flow

When measurements made at CPP < 60 mm Hg were compared with steady state the distribution of regional CBF was considerably different in the balloon group but not so in the infusion group (Fig. 9).

Analysis of CBF dynamics in relation to CPP was performed by forming four CPP classes (> 120,

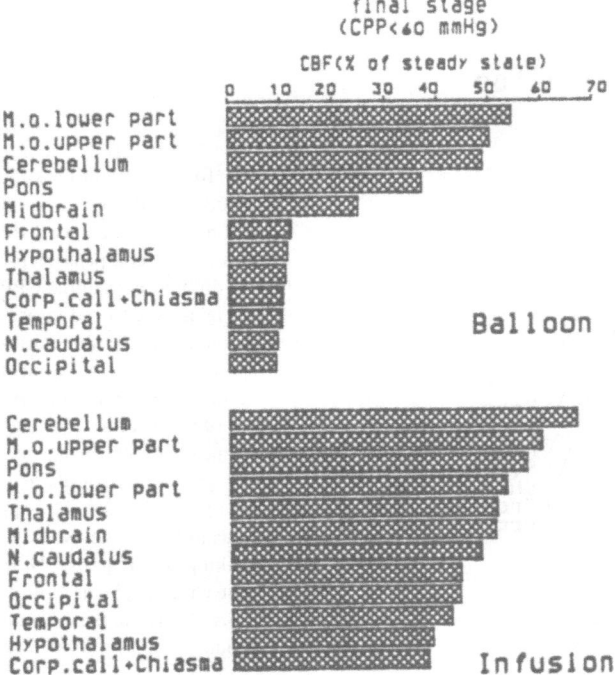

Fig. 8. Relative distribution of flow in different regions after ICP increase and lowering the CPP below 60 mmHg (final stage). Note that medullary, pontine and cerebellar flow are preserved better regardless of the method used to increase the ICP

Table 7. *Maximal pupillary dilatation and midbrain flow.* (N = 7)

Before anisocoria		After anisocoria	
Midbrain flow ml/100 g/min	per cent of control	Midbrain flow ml/100 g/min	per cent of control
23 ± 5.1	60 ± 16	10.3 ± 5.4	24 ± 10

which was forced to pass the point corresponding to a CPP 120 mm Hg, and CBF decrease was expressed in % of this initial value. In spite of the fact that by this construction the profile of CBF change is not represented—this line provided two pieces of information:

1) the mean decrease of CBF over the CPP classes under consideration and

2) the final value of CBF.

This analysis revealed a distinct pattern of regional CBF decrease during balloon expansion (Fig. 10). The regression lines formed 2 distinct groups: The group in which CBF at CPP below 40 was reduced to arround 55% of the initial value and the group with CBF reduction below 15%. The first group comprised lower and upper medulla, cerebellum, pons and the choroid plexus. The second group comprised thalamus, temporal and frontal lobes, caudate nucleus, corpus callosum, and the occipital lobe. The midbrain and hypothalamus showed CBF reduction which could not be related to one of the two groups, but had a tendency towards the group with the lower values. The major

81–120, 41–80, ≤ 40 mm Hg) and calculating the mean of all CBF values in each class.

The relation between CBF means and CPP for each region was roughly approximated by a regression line,

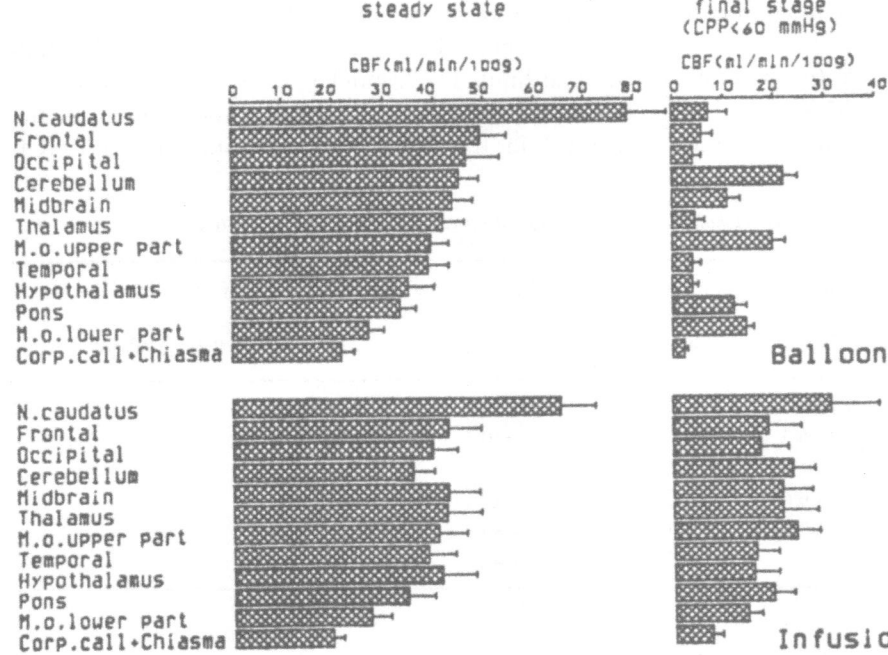

Fig. 9. Distribution of regional blood flow at the steady state and at CPP < 60 mmHg in the balloon and infusion group. Absolute values of flow ± SE are shown. Note the more uniform pattern in the infusion group and the change of flow pattern in the balloon group

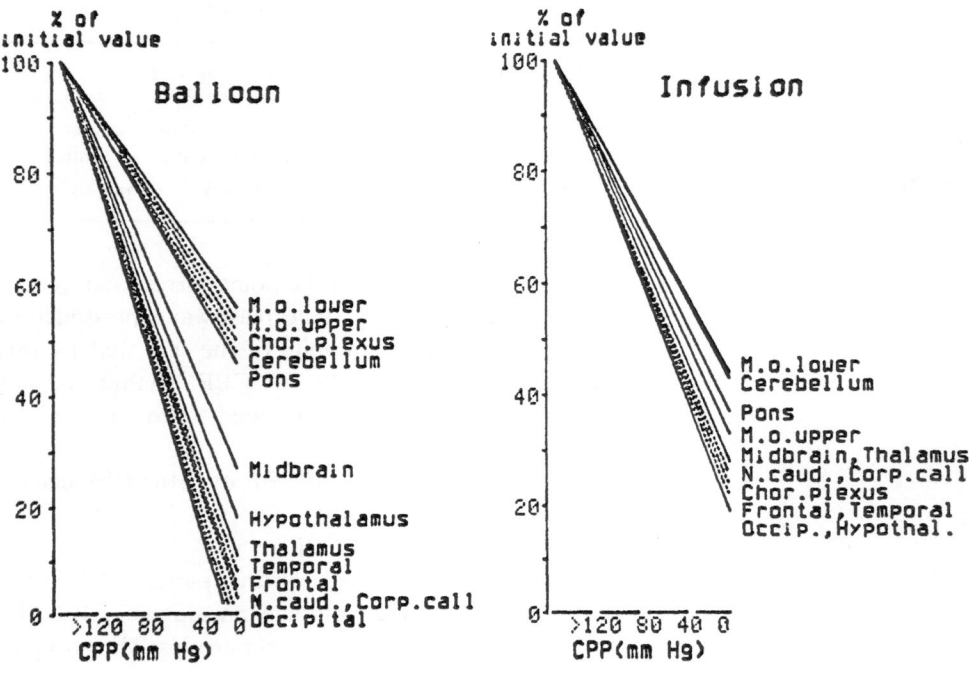

Fig. 10. Regression lines for blood flow decrease for all the regions related to CPP classes. To compensate for the differences in initial CPP the lines were forced to pass through the point corresponding to the control value of the region expressed as 100 per cent. The difference between the clusters of regression lines in the balloon group is statistically significant (p < 0.01). Note the ranking of regions. Note that choroid plexus flow is more affected in the infusion group

difference in the regional blood flow distribution among the animals in the balloon group and in the infusion group was the distinct appearance of blood flow compartmentalization in the first one and lack of this compartmentalization in the second one.

6. Influence of ABP

Between two consecutive CBF measurements spontaneous ABP increase of at least 10% was noted in 9 animals in the balloon group and 8 animals in the infusion group.

ICP was raised by 35 ± 7 mm Hg (mean ± SD) (range 22–58 mm Hg) and 33 ± 12 mm Hg (mean ± SD) (range 20–52 mm Hg) within 4 to 10 and 4 to 18 min respectively. In 10 out of 17 cats from both groups ABP increase took place without or before change of pupillary size whereas in 7 animals vasopressor response occurred at the time when pupillary changes were present.

Differences existed between the influence of ABP increase on CPP and flow depending upon whether blood pressure increase took place before or after pupillary dilatation (Table 8).

Firstly, ABP increase taking place before pupillary dilatation occurs was unlikely to be caused by total brain or medullary ischaemia. CPP was well within normal limits (> 90 mm Hg) in 8 out of 10 cats and CBF and medullary flow were above 70% of the control in the same number. With ABP and ICP

Table 8. *ABP, CPP, CBF and flow in the upper and lower medulla before and during vasopressor response in relation to pupillary dilatation*

	ABP increase without pupillary dilatation (N = 10)		ABP increase with or after pupillary dilatation (N = 7)	
	Values before	Values at max. ABP increase	Values before	Values at max. ABP increase
ABP	138 ± 4.6*	164 ± 4.6	127 ± 6.5	163 ± 12.3
CPP	110 ± 7.8	110 ± 5.3	44 ± 10.7	32 ± 8.0
Blood flow				
BT	63 ± 8.4	43 ± 9.0	10 ± 3.1	7 ± 1.8
MOU	60 ± 18.7	45 ± 9.0	19 ± 5.9	9 ± 3.5
MOL	45 ± 14	34 ± 11	13 ± 4.0	7 ± 2.7

* M ± S.E.

ABP and CPP values in mm Hg. Flow values in ml/min/100 g.

increase change was variable but perfusion was maintained in all animals within normal limits. There was a decrease of total brain flow in 7 animals with flow reduced below 70% of the control only in 3 cases. There was also some reduction of medullary flow again only in 3 animals with flow below 70% of the control. Thus, this type of ABP response seems to be effective in maintaining CPP and blood flow during raising ICP in the majority of cases.

When ABP rose with or after pupillary dilatation the total brain flow and medullary flow was already reduced below 70% of control value in all instances with the exception of a single value for medullary flow in one animal. CPP was < 60 mm Hg in 5 out of 7 animals. Increase of blood pressure would not balance the ICP increase and CPP improved only in one animal. Apart from small variations of 2–3 ml/100 g/min at the level of severe ischaemia, in no instance was an effective improvement of the total or medullary flow observed.

In 5 of 7 animals of this group the EEG tracing was already isoelectric before ABP rise took place and in the remaining two it became isoelectric with a further decrease of flow during the vasopressor response.

Thus, as long as no pupillary signs are present the vasopressor response is not connected with hemisphere or medullary ischaemia and depending on the rate of ICP increase it is able to maintain CPP with only slight reduction of blood flow.

The same pressure response in conjunction with pupillary dilatation gives no net CPP increase and is unable to improve the severely compromised brain and medullary flow.

6.1. Effect of ICP Increase upon Flow in the Spine, Heart and Kidneys

Table 9 shows the values of blood flow in spinal cord, heart and kidneys at steady state and at CPP < 60 mm Hg. The spinal flow decreased during the

experiment in both groups but the pattern of flow change was completely different.

In the infusion group the degree of flow decrease was not different from the change of cerebral blood flow whereas in the balloon group the flow in the spinal cord was much better preserved than in the brain (Fig. 11).

The heart flow behaved differently in the balloon and in the infusion group. In the first one there was a tendency to heart flow reduction and there was a significant (p < 0.01) correlation between the total CBF and the heart flow (Fig. 12).

For the infusion group there was a mean increase of heart flow (from 331 ± 28 ml/min/100 g to 639 ± 78 ml/min/100 g) with much greater variability during the ICP increase. No relation between the CBF and heart flow was found in this group.

The flow in the kidneys decreased in the balloon group but there was no significant change in the infusion group. No relation between CBF change and kidney flow change was found.

6.2. CBF After Cerebral Ischaemia

In 8 animals in the balloon group cerebral ischaemia was maintained for arround 10 min (10.4 ± 2.1 min) with isoelectric EEG trace at ICP levels between 84 and 160 mm Hg. CBF was measured within 5–60 min after ICP release. Sudden ICP release was also produced in 7 animals in the infusion group at ICP levels between 96 and 200 mm Hg. CBF was measured within 6–34 min afterwards.

Table 10 shows the values of ABP, CPP and CBF in animals in which blood flow was measured after ICP release. The balloon and infusion group showed no differences in control values. The minimal CBF during maximal ICP increase and perfusion pressure after ICP release were significantly higher in the infusion group. However, the flow after ICP reduction was increased significantly in comparison with control in the balloon

Table 9. *Flow in the spinal cord and in other organs* (mean ± S.E.M.)

	Balloon		Infusion	
	Steady state (N = 15)	Final stage (Cpp < 60 mm Hg) (N = 13)	Steady state (N = 16)	Final stage (CPP < 60 mm Hg) (N = 8)
Spinal cord (ml/min/100 g)	14.5 ± 1.9	9.6 ± 1.2	14.6 ± 1.8	9.9 ± 1.5
Heart (ml/min/100 g)	285 ± 23	242 ± 23	309 ± 25	699 ± 71
Kidneys (ml/min/100 g)	183 ± 23	131 ± 15	170 ± 15	197 ± 16

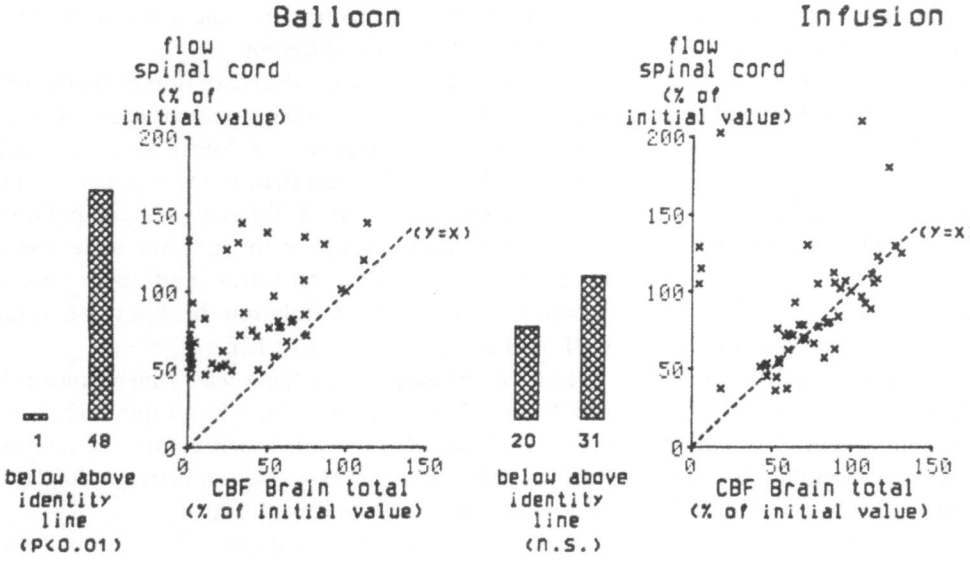

Fig. 11. Spinal cord blood flow change in relation to CBF in the balloon and infusion group. Indirect evidence of compartmentalization

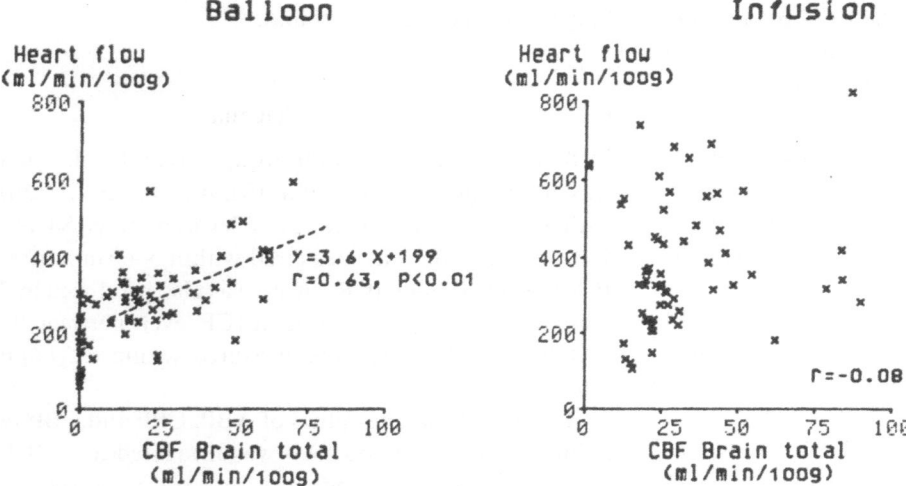

Fig. 12. Heart flow in relation to CBF in the balloon and infusion group. Significant correlation between CBF and heart flow in the balloon group. r = 0.63

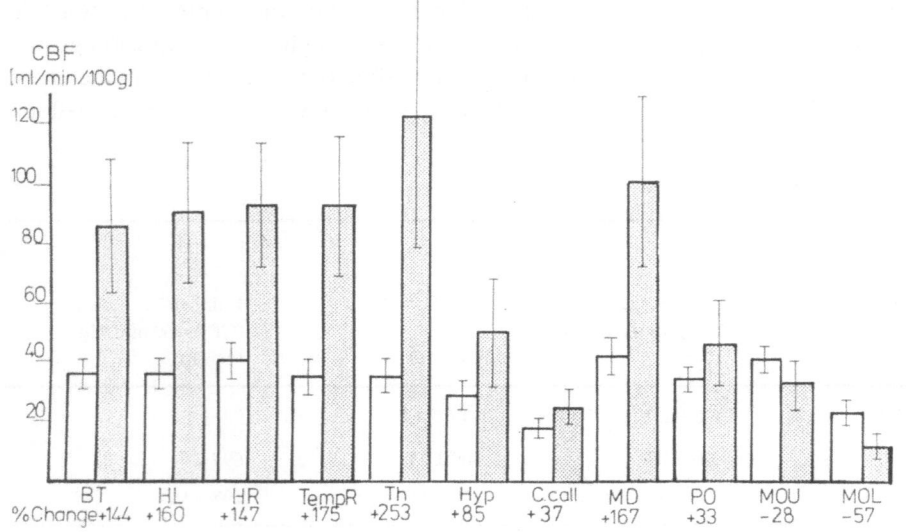

Fig. 13. Absolute values of blood flow in different brain regions after sudden ICP decrease in 8 animals of the balloon group in comparison with control. Note excessive hyperperfusion in the thalamus and midbrain. Values are means ± S.E.

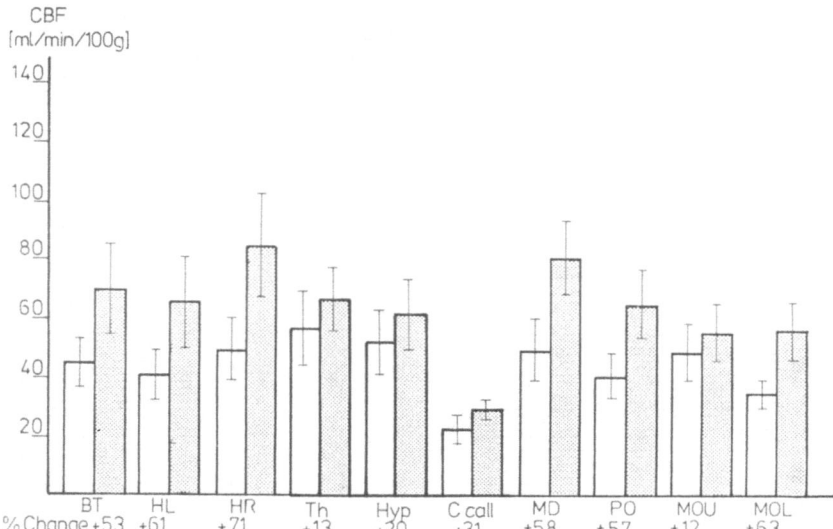

Fig. 14. Absolute values of blood flow in different brain regions after sudden ICP decrease in 7 animals of the infusion group in comparison with control. Note the moderate and more uniform increase of flow when compared with the balloon group

group but not in the infusion group. This is shown for the representative regions in Fig. 13 and Fig. 14.

Thus, the ICP increase of the same magnitude produced different responses depending on the method used. Although perfusion pressure was higher in animals which were subjected to infusion the hyperperfusion was mild and insignificant. In contrast it was excessive in animals in the balloon group with redistribution of flow (Fig. 15) in terms of absolute and relative increase in the thalamus and midbrain and moderate change in lower brain stem. The expansion of the balloon seems therefore to have a more profound

Table 10. *ABP, CPP, CBF in animals in whom blood flow was measured after ICP release.* Values are mean ± S.E.

	Balloon N = 8	Infusion N = 7
Control ABP	133 ± 6.3	139 ± 6.7
Control CPP	127 ± 7.4	133 ± 4.19
Control CBF	36 ± 5.1	45.2 ± 9.2
Maximal ICP	111 ± 12	122 ± 17
Minimal CPP	21 ± 6.5	43 ± 9.0
Minimal CBF	5.8 ± 2.0*	31 ± 12.7
ABP after ICP release	123 ± 9.9*	152 ± 11.2
CPP after ICP release	108 ± 7.1**	143 ± 10.3
CBF after ICP release	84.2 ± 22.0*	69.2 ± 15.2

* Indicates signif. difference in comparison with control, $p < 0.05$. Values for ABP, CPP and ICP in mm Hg. CBF — expressed in ml/min/100 g.

** Indicates signif. difference in comparison with infusion, $p < 0.05$.

effect upon the vasculature than uniform pressure increase caused hydrostatistically.

The degree of impairment of perfusion during ICP increase seems to influence the magnitude of hyperperfusion; there was a significant negative correlation between the perfusion pressure at maximal ICP increase and flow after ICP release in the balloon group ($y = -1.73 \times + 118$; $r = -0.623$, $p < 0.05$). Hyperperfusion after ICP release in the balloon group involved all supratentorial regions with greatest increase in thalamus and midbrain. There was a tendency to flow decrease in the medulla (Fig. 15). In the infusion group the degree of hyperperfusion was more uniform and less for the majority of the regions. Extreme regional hyperperfusion was found in individual animals in the balloon group at the site of compression (Fig. 16).

Discussion

Blood flow values for total brain, heart and kidney in this series are comparable with data reported in the literature and obtained with the same technique[18, 30, 67, 69] but regional flow values differ, which may be explained by differences in sampling and anaesthesia. Comparisons with other techniques like hydrogen clearance or 133 Xenon clearance[28] showed rather variable correlation, but it is agreed that measured flows are dependent on the method used[27] and may be discrepant even in the same animal under comparable experimental conditions[39]. For measurement of flow in multiple areas hydrogen clearance technique which permits unlimited number of measurements can be

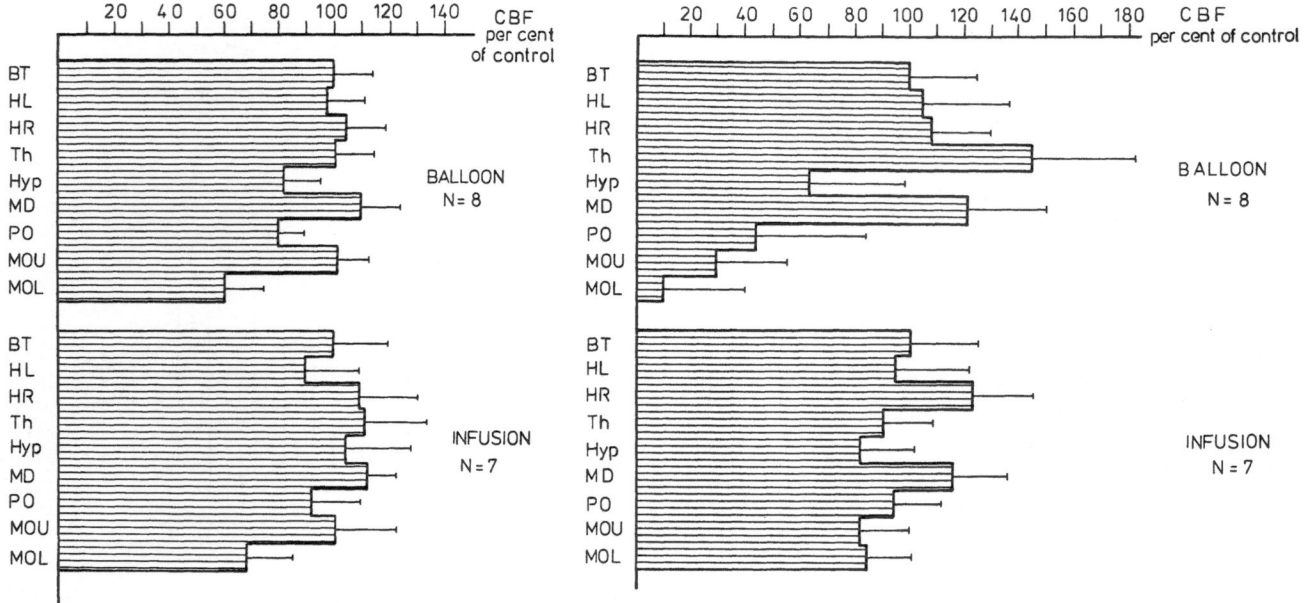

Fig. 15. The profile of blood flow distribution after sudden decrease of ICP in comparison with control. Note the change of blood flow distribution in the balloon group

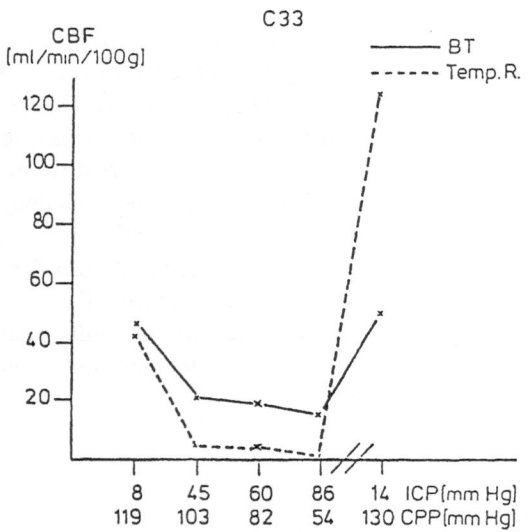

Fig. 16. Relation between total brain blood flow (BT) and regional blood flow at the site of compression in the right temporal region (Temp. R.) during continuous expansion of the balloon. After deflation of the balloon total CBF returns to control value. Hyperperfusion at the site of compression

used[1, 77]. Autoradiography[64] is regarded as most accurate but only one measurement is possible. Microsphere method permits serial intermittent flow estimations at short intervals. The accuracy at the method has been validated[2, 9, 13, 24, 40, 50]. The method provides information about capillary and nutritional flow in morphological terms[26]. The number of capil-laries embolized with the amount of microspheres injected in this series was found to have no influence upon cerebral metabolism, electroencephalogram and neurological condition[40], but with the amount injected cardiovascular dynamics may be affected. The disadvantage lies in poor definition of flow in marginal zones because differences between adjacent small regions in relatively large tissue samples may cancel each other[27].

This study indicates that during continuously increasing ICP produced by focal expanding lesion or by increasing the pressure in CSF spaces without brain herniation the total CBF decreases at CPP levels which are higher than usually reported range observed in studies where stepwise increase of ICP was applied[23, 32, 33, 34, 53, 78].

In spite of adequate CPP, reduction of CBF occurs during initial phase of ICP increase. Lowell and Bloor[47] have made similar observations in experiments with balloon expansion in monkeys and concluded that the inflation of the balloon appeared to lead to loss of autoregulation. Langfitt et al.[41, 43] also found that CBF decreases as soon as ICP rises but stated that rate of pressure increase and deep anaesthesia with barbiturates may have played a role. In later studies[53] immediate CBF reduction was observed only after repeated balloon inflation insults, however, once ICP had started to increase the superimposition of arterial blood pressure in either direction caused a passive pressure/flow relation[45]. Johnston et al.[32, 33, 34] using cisternal infusion and supra- and infratentorial balloon expansion found that immediate CBF decrease was present only with infratentorial expanding lesion.

As pointed out by Meinig et al.[52] and Symon et al.[71] the reported differences in the results of ICP increase upon CBF may be related to different techniques of producing ICP increase, level of ICP, state of vasomotor activity and anatomical species differences particularly of the venous outflow.

In the initial phase of ICP increase the pattern of change of regional CBF was not uniform with sometimes total or regionally different flow changes. Up to the levels of 30–40 mm Hg regulation of flow is effected by decrease of arteriolar resistance[47]. According to this concept during continuous increase of ICP the increase to venous outflow overrides it explaining the apparent lack of flow regulation and the tendency to flow reduction at higher CPP levels. The rate at which ICP increase was produced in our experiments might explain the tendency of CBF decrease in higher CPP classes in both groups.

At the same pressure levels the flow tended to be preserved better when CSF volume was increased than with balloon inflation which confirms the observations of Lewis and McLaurin[46]. Another possible explanation for the difference of CBF/ICP relation between two groups of animals is the increase of ABP.

The concept of perfusion pressure cannot be applied for comparison of different regional flows during continuously increasing ICP connected with brain compression[52]. Different regional flows are connected with differences of regional or local perfusion pressure at precapillary and capillary level. There is evidence that in a given compartment tissue pressure gradients are present only during volume increase and that the gradient capillary/local brain tissue pressure is the main controlling factor of the flow[5, 6]. In our series it was not possible to measure directly the tissue pressure in regions in which CBF was studied. The "cotton wick" method used for this purpose was applied for slight increases of ICP[60, 74] and fails at higher levels and long-term measurement[12]. To evaluate regional change to flow during progressive brain compression and eliminate this problem flows between the regions were compared directly expressing the blood flow in per cent of its own control value. Differences were noted both in the rate of change and in the levels of flow depression observed in different parts of the brain in the balloon group. This is almost certainly related to the effects of brain distortion and strongly supports the concept of pressure gradients developing across parts of the brain

tissue and the hypothesis of decisive role of local pressure in influencing the flow[51]. As expansion of the balloon causes both interhemispheric and craniocaudal pressure differences[4] and tentorial impaction in cats may occur at pressures below 25 mm Hg[51], we believe that interregional changes in blood flow may reflect regional pressure gradients.

Tentorial herniation leads to distinct compartmentalization with significant difference in the degree and rate of CBF change between the hemispheric structures and the brain stem. In the lower brain stem (pons and medulla oblongata) the reduction of CBF is smaller and less rapid than in other regions[66, 67]. In another series of experiments in our laboratory morphological changes occurring with repeated ICP increase produced in the same manner as in this study were analysed[52].

Two zones of primary ischaemia were found in cats subjected to balloon compression: the first passed through the gyrus cinguli-thalamus-corpora mamillaria; the second zone run from inferior and superior colliculi through the tegmentum to the border of the pons. The ischaemic lesions in the thalamo-hypothalamic areas appeared earlier than in the mesencephalic-pontine region. This would fit with findings indicating different dynamics of CBF reduction between thalamus, midbrain and pons, although this is still only a rough reflection of complex chain of events occurring at the incisural hiatus during the process of herniation.

We were unable to disclose side differences in flow reduction in the midbrain, pons and medulla during continuous inflation of the balloon. This contrasts with the asymmetry in flow reduction in supratentorial structures. Anisocoria occurred when midbrain flow was reduced to about 60% of control flow, but it appears that during progression of pupillary dilatation the midbrain flow is greatly reduced.

Sunami et al.[70] who investigated the flow in thalamus, inferior colliculus and medulla oblongata with hydrogen clearance method found that flow in the inferior colliculus and medulla did not decrease until infratentorial pressure exceeded 20 and 40 mm Hg respectively and that the flow in the midbrain was markedly reduced whereas the flow in the medulla did not decrease until further introduction of volume.

Different dynamics of CBF reduction revealed in central structures with a pattern of deterioration along the action force vectors in rostro-caudal direction supports the observations that pressure gradients develop during the critical phase of ICP increase[71].

Compartmentalization of CSF spaces and pressures caused by supratentorial compression is reflected by significant interregional differences of CBF.

In particular the flow in the brain stem is preserved better than in supratentorial regions. Malik et al.[49] speculated that the pressure gradient might account for preservation of blood flow to medulla.

This study however shows clearly that redistribution of blood flow during intracranial hypertension is not only the result of heterogenous increases in ICP. *Independently from the method used to increase the ICP it seems that the brain stem compensates better for global CPP reduction than supratentorial regions resulting in better preservation of flow.* This is supported by similar observations of Sadoshima et al.[63] who used stepwise infusion of mock-CSF. The same type of redistribution of CBF was found when perfusion pressure was reduced by hypotension[54].

Our results indicate that flow at CPP below 60 mm Hg was more affected by ICP increase produced by expansion of the balloon than when infusion was used. Possibly, the blood pressure change during the experiment influenced this result. As pointed out by Grubb et al.[22] and Leech and Miller[45] the resistance of flow to ICP increase is better than to decrease of blood pressure. In the balloon group, additionally to CPP reduction through ICP increase a decrease of ABP was present. This tendency has been demonstrated in experiments with balloon inflation by Fitch et al.[16, 17]. In contrast in the infusion group there was an increase of ABP which may be one of the reasons for the higher average flow at the final stage in the latter group. This could be explained by large systemic volume effect or direct pressure effect upon the spinal cord[56]. In spite of the different ABP reaction during increasing ICP between both groups the ranked blood flow reduction was similar (see Fig. 8).

ABP reaction could also lead to differences in haemodynamic demonstrated by the change in the heart flow and kidney flow. The correlation between the cerebral flow and heart flow which appeared when ICP was increased by balloon inflation is intriguing but it may be coincidental and requires further study including the variation of the cardiac output. It is not quite clear in what way the CBF reacts to changes in cardiac output. Davis and Sundt[14] and Pearce and D'Alcey[57] demonstrated a significant decrease of CBF in cats and in dogs with moderate decrease of cardiac output and hypovolemia without changes in blood pressure. This is at variance with the current concept of autoregulation[23, 59].

There are very few data available about the influence of ICP upon the cardiac output/CBF relationship. A complex pattern of organ flow distribution change and cardiac output increase was found when ICP was raised by cisternal infusion in lambs[38]. In brain trauma model immediate and lasting cardiac output and stroke volume reduction was found but when ICP increase was produced by inflation of the balloon cardiac output was reduced only when slowing of the heart rate was concomitant[7]. Brain stem suffered least during arterial hypotension in comparison to cerebellar of cerebral white matter[35]. During haemorrhagic hypotension in dogs a significant redistribution of cerebral blood flow was found to take place in the thalamus, mesencephalon, pons and medulla with the elevation of rCBF/cardiac output ratio in these regions[10, 18, 35] but discrepancies between species were reported[19, 29]. The mechanism of preferential redistribution of cardiac output in the brain at late stages of haemorrhagic hypotension remains unknown but there is a striking similarity between the above findings and redistribution of blood flow during increasing ICP indicating that redistribution of cerebral blood flow favours those areas where neurons related to cardiovascular control are located. In our series of experiments no significant redistribution toward diencephalic structures (thalamus and hypothalamus) was observed but redistribution of flow towards medulla and pons was quite clear. Haemorrhage and hypotension cause excitation of neurons responsible for cardiovascular regulation by changes in afferent impulses from reflexogenic areas in the circulation[11] and by release of vasoactive substances. Excitation of neuronal activity in the midbrain and pons was also noted during initial ICP increase by George[20] and was followed by progressive, rostrocaudal paralysis of neuronal activity. This coincides with our findings of flow increase in caudal brain stem observed in several animals in our series and was further supported by study of BAEP during ICP increase produced by balloon inflation in cat where a brief period of increase of the amplitude of evoked potentials preceded the reduction, deformation, latency increase and finally complete loss at high ICP level[36].

Hyperperfusion is a common physiological reaction to even short ischaemia. Studies on postischaemic brain circulation showed heterogeneous flow patterns after ischaemia[20, 73]. We use the term hyperperfusion as no measurements of transit time and cerebral blood volume were made which would justify the term hyperemia[73]. Also the term luxury-perfusion[44] should not be used unless reduced O_2 or glucose consumption are proven. Hossmann et al.[30] reported above 500% increase of flow in the brain stem and basal ganglia following 1 hour of complete ischaemia and explained the change of the rank order of regional flows as the result of regional change of the number of vessels perfused. These authors found no postischaemic shunting.

ICP increase to comparable levels followed by ICP release had a different effect on postischaemic CBF. During maximal ICP increase the flow was preserved better in animals in the infusion group with correspondingly less hyperperfusion when ICP was returned to normal. Distortion of the vasculature followed by vasoparalysis is the likely factor responsible for these differences between the balloon and infusion group.

The degree and duration of ischaemia produced by increased ICP influence the postischaemic hyperperfusion. This is of importance considering that hyperperfusion following ischaemic episodes is thought to be responsible for enhancement of progressive postischaemic oedema[15, 43, 78]. As there is less ischaemia at the lower brain stem at the time when the flow ceased in the hemispheres no hyperperfusion and slight tendency to reduction of flow occurs at medullary level. This is likely to be due to redistribution of the flow. Hyperperfusion in the ischaemic thalamus and midbrain would also explain the preferential zone of early blood-brain barrier damage found in front of and behind the tentorium.

In experimental animals expansion of supratentorial mass sufficient to cause transtentorial herniation always produced caudal displacement and distortion of the brain stem. Both mechanisms—lateral compression of the brain stem[26], axial displacement[72] and longitudinal buckling[31]—account for clinical symptoms. The axis of the brain stem in cat is different from man and this animal is more likely to develop axial displacement. Apart from rostro-caudal deterioration of flow disclosed in our experimental group we were not able to demonstrate side differences in the brain itself, but the method used for flow estimations is not suitable for study of flow in very small areas. Caudal displacement may be sufficient to explain the vasopressor response through direct mechanical effect upon medullary centers without need of medullary ischaemia.

The initiation of the vasopressor response is, however, still a matter of dispute. Misu et al.[55] found unexpectedly early transmission of increased ICP from the supratentorial space to the brain stem and recorded the vasopressor response in monkeys when the midbrain begun to shift dorsally. Schrader[66, 67] considers brain stem distortion per se as an unlikely mechanism to elicit the vasopressor response and postulates ischaemia in the brain stem as the probable mechanism. Using the radioactive microsphere method for blood flow estimation of several organs he concluded that the sole mechanism mediating the systemic hypertensive response is the peripheral vasoconstriction causing reduced blood flow in all of the extracellular organs except for heart muscle and the adrenal medulla. The CBF threshold for appearance of SEP latency changes were 15–20 ml/100 g/min and for BAEP changes less than 15 ml/100 g/min[37], which confirms that the effect of ischaemia due to rised ICP appears earlier in the white matter of the hemispheres than in the brain stem[3].

Koehler et al.[37] who studied the regional blood flow with microsphere method during diffuse ICP increase in sheep found, that CBF is maintained down to CPP of 40–50 mm Hg. With CBF reduction oxygen extraction rises and the uptake is maintained unless CBF is reduced more than 30–40%. They found that pressure response permits to preserve some degree of CPP but is not adequate to restore the oxygen uptake or to improve the somatosensory of brain acoustic evoked potentials.

Our results concerning the compartmentalization of flow and sparing of flow in the lower brain stem until the final stages of increased ICP have been recently confirmed by Schrader et al.[67]. In our series, however, vasopressor response at a moment when still no pupillary dilatation was present was not connected with cerebral ischaemia. It maintains CPP but does not improve the flow[17, 61]. Hypertensive response with or after pupillary dilatation does not help to improve already profoundly disturbed blood flow. This confirms the earlier observations of Shalit and Cotev[68].

Acknowledgements

My particular thanks are due to Dr. Hoffmann, Dr. Kurzaj, Dr. Winkler and to Mr. Buss and Mr. Anderl for their invaluable help and advice while performing this study.

References

1. Aukland K, Bower BF, Berliner RW (1964) Measurement of local blood flow with hydrogen gas. Circ Res 14: 164–187
2. Baethmann A (1986) Blood flow and metabolism in brain stem—pathophysiological implications. In: Samii M (ed) Surgery in and around the brain stem and the third ventricle. Springer, Berlin Heidelberg New York, pp 95–101
3. Branston NM, Ladds A, Symon L, Wang AD (1984) Comparison of the effects of ischaemia on early components of the

somatosensory evoked potential in brain stem, thalamus and cerebral cortex. J Cereb Blood Flow Metab 4: 68–81

4. Brawanski A, Gaab MR (1981) Intracranial pressure gradients in the presence of various intracranial space occupying lesions. In: Schiefer W, Klinger M, Brock M (eds) Advances in neurosurgery 9. Springer, Berlin Heidelberg New York, pp 355–362

5. Brock M, Furuse M, Weber R, Hasuo M, Dietz H (1975) Brain tissue pressure gradients. In: Lundberg N, Ponten U, Brock M (eds) Intracranial pressure II. Springer, Berlin Heidelberg New York, pp 215–217

6. Brock M, Hadjidimos AA *et al* (1971) The effects of hyperventilation on regional cerebral blood flow. On the role of changes in intracranial pressure and tissue perfusion pressure for shifts in rCBF distribution. In: Toole JF, Moossy J, Janeway R (eds) Cerebral vascular disease. Grune & Stratton, New York London, pp 114–123

7. Brown FD, Velasquez LS, Lin CY, Johns L, Mullan S (1983) Cardiovascular effects of a graded increase in ICP. In: Ishii S, Nagai H, Brock M (eds) Intracranial pressure V. Springer, Berlin Heidelberg New York, pp 463–467

8. Buckberg GD, Luck JC, Payne DB, Hoffman JIE, Archie JP, Fixler DF (1971) Some sources of error in measuring regional blood flow with radioactive microspheres. J Appl Physiol 31: 598–604

9. Chen RY, Fan FC *et al* (1983) Effects of sphere size and injection site on regional cerebral blood flow measurements. Stroke 14: 769–776

10. Chen RY, Fan FC *et al* (1984) Regional cerebral blood flow and oxygen consumption of the canine brain during haemorrhagic hypotension. Stroke 15: 343–350

11. Chien S (1967) Role of sympathetic nervous system in haemorrhage. Physiol Rev 47: 214–288

12. Clark RM, Capra NF, Halsey JH (1976) Method for measuring brain tissue pressure: Response of alteration of pCO_2 systemic blood pressure and middle cerebral artery occlusion. J Neurosurg 43: 1–8

13. Dahners H, Flohr H, Meyer M, Christ R (1970) Messung der peripheren Verteilung des Herzzeitvolumens mit radioaktiv markierten Partikeln. Pflügers Archiv 319: R 28–R 29

14. Davis D, Sundt TM (1980) Relationship of cerebral blood flow to cardiac output, mean arterial pressure, blood volume and alpha and beta blockade in cats. J Neurosurg 52: 745–754

15. Durward QY, del Maestro RF, Amacher AL, Farrar JK (1983) The influence of systemic arterial pressure and intracranial pressure on the development of cerebral vasogenic edema. J Neurosurg 59: 803–809

16. Fitch W, McDowall DG, Keaney NP, Pickerodt WA (1977 a) Systemic vascular responses to increased intracranial pressure. 1. Effects of progressive epidural balloon expansion on intracranial pressure and systemic circulation. J Neurol Neurosurg Psychiat 40: 833–842

17. Fitch W, McDowall DG, Keaney NP, Pickerodt WA (1977 b) Systemic vascular responses to increased intracranial pressure. 2. The "Cushing" response in the presence of intracranial space-occupying lesions, systemic and cerebral haemodynamic studies in the dog and the balloon. J Neurol Neurosurg Psychiat 40: 843–852

18. Fritschka E, Artigas J, Shigeno T, Minguillon C, Cervós-Navarro J (1981) Regional cerebral blood flow after occlusion of the middle cerebral artery in cats, correlation of blood flow and specific gravity at 24–48 hours post occlusion. In: Cervós-

Navarro J, Fritschka E (eds) Cerebral microcirculation and metabolism. Raven Press, New York, pp 433–441

19. Gamache FW, Myers RE, Monell E (1976) Changes in local cerebral blood flow following profound systemic hypotension. J Neurosurg 44: 215–225

20. George B (1980) Neurophysiological effects of experimental intracranial hypertension on three different structures of the brain stem in the cat. Rostro-caudal deterioration. Acta Neurochir (Wien) 55: 71–83

21. Ginsberg MD, Budd WW, Welsh FA (1978) Diffuse cerebral ischemia in the cat. I. Local blood flow during severe ischaemia and recirculation. Ann Neurol 3: 482–492

22. Grubb RL, Raichle ME, Phelps ME, Rutcheson RA (1975) Effects of increased intracranial pressure on cerebral blood volume, blood flow and oxygen utilization in monkeys. J Neurosurg 43: 385–398

23. Häggendal E, Löfgren J, Nilsson NJ, Zwetnow NN (1970) Effects of varied cerebrospinal fluid pressure on cerebral blood flow in dogs. Acta Physiol Scand 79: 262–271

24. Hales JRS (1973) Radioactive microsphere measurement of cardiac output and regional blood flow in the sheep. Pflügers Arch 344: 119–132

25. Hales JRS, Yeo JD, Stabback S, Fawcett AA, Kearns R (1981) Effects of anesthesia and laminectomy on regional spinal cord blood flow in conscious sleep. J Neurosurg 54: 620–626

26. Hasenjäger T, Spatz H (1937) Über örtliche Veränderungen der Konfiguration des Gehirns beim Hirndruck. Arch Psychiat 107: 193–222

27. Heiss WD (1981) Editorial: What method to choose for qualification of cerebral blood flow in experimental research. Stroke 12: 555–556

28. Heiss WD, Traupe H (1981) Comparison between hydrogen clearance and microsphere technique for rCBF measurement. Stroke 12: 161–167

29. Heistad DD, Marcus ML, Gross PM (1978) Effects of sympathetic nerves on cerebral vessels in dog, rat, and monkey. Am J Physiol 235: 4544–4552

30. Hossmann KA, Hossmann V, Takagi S (1978) Microsphere analysis of local cerebral and extracerebral flow after complete ischaemia of the cat brain for one hour. J Neurol 218: 275–285

31. Howell DA (1961) Longitudinal brain stem compression with buckling. Arch Neurol Psychiat 4: 116–123

32. Johnston IH, Rowan JO, Harper AM, Jennet WB (1972) Raised intracranial pressure and cerebral blood flow. 1. Cisterna magna infusion in primates. J Neurol Neurosurg Psychiat 35: 285–296

33. Johnston IH, Rowan JO (1974) Raised intracranial pressure and cerebral blood flow. 4. Intracranial pressure gradients and regional cerebral blood flow. J Neurol Neurosurg Psychiat 37: 585–592

34. Johnston IH, Rowan JO, Harper AM, Jennet WB (1973) Raised intracranial pressure and cerebral blood flow. 2. Supra- and infratentorial mass lesions in primates. J Neurol Neurosurg Psychiat 36: 161–170

35. Kassell NF, Boazim DJ, Olin JJ, Sprowell JA (1983) Cerebral and systemic circulatory effects of arterial hypertension induced by adenosine. J Neurosurg 58: 69–76

36. Klug N (1984) Funktionsuntersuchungen des Hirnstamms im akuten Mittelhirnsyndrom unter Berücksichtigung vegetativer Meßgrößen während der Decerebration. Thesis, Giessen

37. Koehler RC, Backofen JF, McPherson RW, Rogers MC, Traystman RJ (1986) Relationship of regional cerebral blood

flow, evoked potential responses, and systemic hemodynamics during intracranial hypertension. In: Miller JP, Teasdale GM, Rowan JO, Galbraith SL, Mendelow AD (eds) Intracranial pressure VI. Springer, Berlin Heidelberg New York, pp 365–368

38. Koehler RC, Backofen JE, Traystman RJ, Jones MD Jr, Rogers MC (1983) Peripheral organ blood flow distribution during raised intracranial pressure in young lambs. In: Ishii S, Nagai H, Brock M (eds) Intracranial pressure V. Springer, Berlin Heidelberg New York Tokyo, pp 880–884

39. Lacombe P, Meric P, Seyhaz J (1980) Validity of cerebral blood flow measurements obtained with quantitative tracer techniques. Brain Res Rev 2: 105–169

40. La Morgese J, Fein IM, Shulman K (1975) Polarographic and microsphere analysis of ultraregional cerebral blood flow rates in cats. In: Harper M *et al* (eds) Blood flow and metabolism in the brain. Churchill & Livingstone, New York Edinburgh London, pp 73–79

41. Langfitt TW, Kassel NT, Weinstein JD (1965) Cerebral blood flow with intracranial hypertension. Neurology 15: 761–773

42. Langfitt TW, Weinstein JD, Kassel NF *et al* (1964) Transmission of increased intracranial pressure. I. Within the craniospinal axis. J Neurosurg 21: 989–997

43. Langfitt TW, Weinstein JD, Kassel NF (1965) Cerebral vasomotor paralysis produced by intracranial hypertension. Neurology (Minn) 15: 622–641

44. Lassen NA (1966) The luxury perfusion syndrome and its possible relation to acute metabolic acidosis localized within the brain. Lancet 2: 1113–1116

45. Leech P, Miller JD (1974) Intracranial volume-pressure relationships during experimental brain compression in primates. II. Effect of induced changes in systemic arterial pressure and cerebral blood flow. J Neurol Neurosurg Psychiat 34: 1099–1104

46. Lewis HP, McLaurin RL (1971) Regional cerebral blood flow and increased intracranial pressure produced by simulated hydrocephalus, mass lesion and cerebral edema. Surg Forum 22: 424–426

47. Lowell HM, Bloor BM (1971) The effect of increased intracranial pressure on cerebrovascular haemodynamics. J Neurosurg 34: 760–769

48. Makowski EL, Meschia G, Dwegemueller W, Battagia FC (1968) Measurement of umbilical arterial blood flow to the sheep placenta and fetus in utero. Circ Res 23: 623–631

49. Malik AB, Krasney JA, Proyce GJ (1977) Respiratory influence on the total and regional cerebral blood flow responses to intracranial hypertension. Stroke 8: 243–249

50. Marcus ML, Heistad DP *et al* (1974) Total and regional cerebral blood flow measurement with 7–10, 15-, 25-, and 50 μm microspheres. J Appl Physiol 40: 501–507

51. Marmarou A, Takagi H, Walstra G, Shulman K (1980) The role of brain tissue pressure in autoregulation of CBF in areas of brain edema. In: Shulman K, Marmarou A, Miller JD, Becker DP, Hochwald GM, Brock M (eds) Intracranial pressure IV. Springer, Berlin Heidelberg New York, pp 257–260

52. Meinig G, Reulen HJ, Magalvy C, Hase U, Hey O (1972) Changes of cerebral hemodynamics and energy metabolism during increased CSF pressure and brain edema. In: Brock M, Dietz H (eds) Intracranial pressure I. Springer, Berlin Heidelberg New York, pp 79–84

53. Miller JD, Stanek A, Langfitt TW (1971) Concepts of cerebral perfusion pressure and vascular compression during intracranial hypertension. Progr Brain Res 35: 411–432

54. Miller JD, Stanek AE, Langfitt TW (1973) Cerebral blood flow regulation during experimental brain compression. J Neurosurg 39: 186–196

55. Misu N, Kuchiwaki M, Hirai N, Ishiguri H, Takada S, Kageyama N (1986) Local shift of the brain and its relation to the tentorial edge. In: Miller JD, Teasdale GM, Rowan JO, Galbraith SL, Mendelow AD (eds) Sutracranial pressure VI. Springer, Berlin Heidelberg New York, pp 318–324

56. Pásztor A, Pásztor E (1980) Spinal vasomotor reflex and Cushing response. Acta Neurochir (Wien) 52: 85–97

57. Pearce WJ, d'Alecy LG (1977) Normotensive haemorrhage and cerebral blood flow. Acta Neurol Scand [Suppl 56] 64: 334–335

58. Rap ZM, Csécsei G, Klug N, Zierski J, Pia HW (1983) Morphological analysis of experimental decerebration after acute epidural compression. In: Auer LM, Loew F (eds) The cerebral veins. Springer, Wien New York, pp 293–298

59. Rapela CE, Green HD (1964) Autoregulation of canine cerebral blood flow. Circ Res [Suppl 1] 15: 205–211

60. Reulen HJ, Graham R, Spatz M, Klatzo I (1977) Role of pressure gradients and bulk flow dynamics of vasogenic brain edema. J Neurosurg 46: 24–35

61. Rowan JO, Teasdale G (1977) Brain stem blood flow during raised intracranial pressure. In: Ingvar DH, Lassen NA (eds) Cerebral function, metabolism and circulation. Munksgaard Ltd, Copenhagen, pp 520–521

62. Rudolph AM, Heymann AM (1967) The circulation of the fetus in utero. Circ Res 21: 163–184

63. Sadoshima S, Thames M, Heistad D (1981) Cerebral blood flow during elevation of intracranial pressure. Role of sympathetic nerves. Am J Physiol 241: H 178–H 184

64. Sakurada O, Kennedy C, Jehle J, Brown JD, Carbin GL, Sokoloff L (1978) Measurement of local cerebral blood flow with iodo (^{14}C) antipyrine. Am J Physiol 234: H 59–H 66

65. Schaper W, Lewi P, Flameng W, Gijpen L (1973) Myocardial steal produced by coronary vasodilatation in chronic coronary artery occlusion. Basic Res Cardiol 68: 3–20

66. Schrader H (1985) Dynamics of intracranial expanding masses. A/S. Holshed-Trykk, Oslo, pp 7–20

67. Schrader H, Zwetnow NN, Löfgren J, Hall C, Mørkrid L (1986) Mechanism of the Cushing response. In: Miller JD, Teasdale GM, Rowan JO, Galbraith SL, Mendelow AD (eds) Intracranial pressure VI. Springer, Berlin Heidelberg New York, pp 378–384

68. Shalit MN, Cotev S (1975) The "Cushing response". A compensatory mechanism or a dangerous phenomenon? In: Lundberg N, Ponten U, Brock M (eds) Intracranial pressure II. Springer, Berlin Heidelberg New York, pp 307–310

69. Shulman K, Furman H, Rosende R (1973) Regional cerebral blood flow—evaluation of the microsphere technique. Stroke 4: 338–342

70. Sunami N, Tsutsui T, Houma Y, Fujimoto S, Nagao S, Ohmoto T, Nishimoto A (1983) Changes in local blood flow of the brain stem in acute intracranial hypertension. In: Ishii S, Nagai H, Brock M (eds) Intracranial pressure V. Springer, Berlin Heidelberg New York Tokyo, pp 458–462

71. Symon L, Pásztor E, Branston MN, Dorsch WC (1974) Effect of supratentorial space-occupying lesions on regional intracranial pressure and local cerebral blood flow: an experimental study in baboons. J Neural Neurosurg Psychiat 37: 617–626

72. Thompson RK, Malina S (1959) Dynamic axial brain stem distortion as a mechanism explaining the cardio-respiratory

changes in increased intracranial pressure. J Neurosurg 16: 664–675

73. Traupe H, Kruse E, Heiss WD (1982) Reperfusion of focal ischemic of varying duration. Postischemic hyper- and hypoperfusion. Stroke 13: 615–622

74. Tulleken CAF, Meyer JS *et al* (1975) Brain tissue pressure gradients in experimental infarction recorded by multiple wick-type transducers. In: Lundberg N, Ponten U, Brock M (eds) Intracranial pressure II. Springer, Berlin Heidelberg New York, pp 224–227

75. Winkler B (1979) The tracermicrosphere method. In: Schaper W (ed) The pathophysiology of myocardial perfusion. Elsevier/North Holland Biomedical Press, Amsterdam New York Oxford, pp 13–42

76. Winkler B, Stämmler G (1982) Measurement of radioactive tracer microsphere blood flow with NaI (TI)- and Ge-well-type detectors. Belgique Research in Cardiology 77: 292–300

77. Young W (1980) H_2 clearance measurement of blood flow: a review of technique and polarographic principles. Stroke 2: 552–564

78. Zwetnow NN (1970) Effects of increased cerebrospinal fluid pressure on the blood flow and on energy metabolism of the brain. An experimental study. Acta Physiol Scand [Suppl 339] 79: 1–31

Author's address: Prof. Dr. Jan Zierski, Department of Neurosurgery, Hospital Neukoelln, Rudower Strasse 48, D-1000 Berlin 47.

Acta Neurochirurgica, Suppl. 40, 117–130 (1987)

Biomathematics of Intracranial CSF and Haemodynamics. Simulation and Analysis with the Aid of a Mathematical Model

Oskar Hoffmann

Department of Neurosurgery, University of Giessen, Federal Republic of Germany

Contents

Summary

A mathematical model of the isolated intracranial system including autoregulation of cerebral blood flow with the aid of a variable cerebrovascular resistance is described. The rate of formation of cerebrospinal fluid is assumed to depend on the regional blood flow through the choroid plexuses. This model is extended by cardiovascular components including the left ventricle of the heart, the aorta and the peripheral resistance. Additionally the model contains control circuits to simulate the short-time behaviour of the blood pressure regulation with the aid of the baroreceptor reflex. Disturbances of central regulation of blood pressure are simulated depending on changes of the regional blood flow through the brain stem.

The application of the model is demonstrated by the analysis of the influence of arterial blood pressure upon the intracranial pulse pressure relationship (PPR) and upon the pressure response to a volume pressure test. Theoretical considerations and simulations reveal an opposite effect of arterial blood pressure (ABP) and its amplitude upon PPR. The ICP amplitude rises with decreasing ABP or increasing ABP amplitude. Breakpoints and other deviations from a linear PPR over the whole ICP range are studied by the analysis of the transfer function.

The application of the model concerning parameter estimation methods is demonstrated and discussed. Simulations of rhythmic phenomena with the aid of the extended model point out possible approaches to quantitative descriptions of disturbances of central regulation.

Keywords: Intracranial pressure; mathematical model; cerebrospinal fluid dynamics.

Introduction

The advantage of model building and simulation has been generally accepted in a wide range of experimental sciences. In contrast to animal models frequently used in medicine physical and mathematical models provide the possibility of easy manipulation of system parameters. Since the application of a physical model is often limited to a particular question and a change of its structure is practically impossible, mathematical models seem to be more efficient[6, 13, 34].

A mathematical model describes the main relationships between the variables of the real system by appropriate equations, usually difference or differential equation. One of the major difficulties in modeling in biomedicine is the high degree of complexity of the systems involving a large number of internal elements and in particular nonlinear interactions[21, 31–33]. The complexity of the model structure should be limited by the purpose for which the model is developed. By overcomplication a model may become unmanageable, even with modern computer technology, and the opportunity to obtain valuable contribution to the understanding of the investigated system may be missed[44].

During the past sixteen years various biophysical and mathematical models of cerebro spinal fluid (CSF) dynamics have been developed[1, 2, 4, 5, 15–18, 30, 36]. Some of these models aimed to give a better insight into the dynamic interactions in the intracranial system, in particular with respect to the pathogenesis of hydrocephalus. Other models have been applied for experimental calculation of static and dynamic parameters of the intracranial system. However, none of these models includes the effects of haemodynamics, in particular the autoregulation of cerebral blood flow, upon CSF dynamics, and vice versa the effect of an increased ICP upon control of cardiovascular is omitted in all models.

The objective of this study is a qualitative and quantitative extension of the known models of intracranial CSF and haemodynamics including new variables and physiological control circuits, in particular:

— autoregulation of cerebral blood flow by the aid of a variable cerebrovascular resistance,

— CSF formation rate as a function of the regional blood flow in the choroid plexuses,

— control of arterial blood pressure via heart rate and peripheral resistance,

— disturbances of this control circuit related to changes of regional blood flow in the brain stem.

Model Equations

1. Intracranial System

The intracranial space is modeled as a rigid container filled with brain tissue, cerebrospinal fluid and blood in the arterial and venous part of the cerebral vascular bed (Fig. 1). Arterial and venous part of the vascular bed

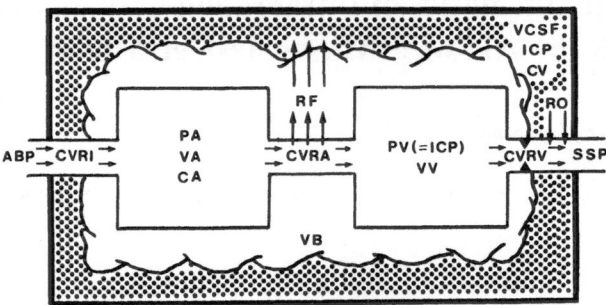

Fig. 1. Model of the isolated intracranial system for simulation of CSF and haemodynamics. Abbreviations are explained in the text

are connected by a vessel with variable flow resistance (CVRA), representing small arteries, arterioles and capillaries. For the purpose of autoregulation of the cerebral blood flow over the range from 60 to 130 mm Hg of cerebral perfusion pressure (CPP) CVRA must vary between 0.06 and 0.15 mm Hg · min/ml (Fig. 2). When CPP drops below 60 mm Hg autoregulation fails. Because of the compression of cerebral vessels coupled with decreasing perfusion pressure the cerebrovascular resistance in-

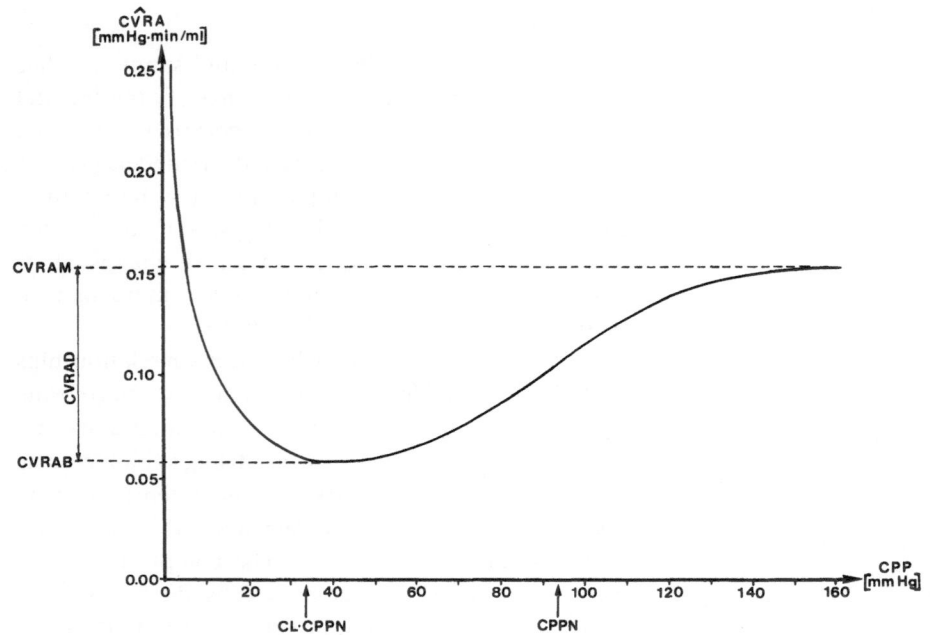

Fig. 2. Cerebrovascular resistance feedback loop for autoregulation of cerebral blood flow. Arterial part of the cerbrovascular resistance (CVRA) as a function of cerebral perfusion pressure (CPP). Rapid increase of CVRA when perfusion pressure drops below the limit CL · CPPN

creases rapidly. Mathematically this can be expressed by

$$
C\widehat{VR}A = \begin{cases} CVRAM - \dfrac{CVRAD}{1 + (CPP/CPPN)^6} \dfrac{CPP}{CPPN} \geq CL \\[4mm] CVRAB \cdot (CK \cdot CPPN/CPP)^2 \quad \dfrac{CPP}{CPPN} < CL \ . \end{cases}
$$

This choice of the feedback loop of autoregulation of cerebral blood flow is in accordance with experimental results reported by Johnston and Rowan[22] and by Lowell and Bloor[29]. Since a change of CPP causes a delayed reaction of CVRA[26] the dynamic behaviour of CVRA is described by the differential equation

$$
\frac{dCVRA}{dt} = -\frac{1}{TAR}CVRA + \frac{1}{TAR}C\widehat{VR}A \ .
$$

The time constant $TAR = 10$ sec provides that an autoregulatory process will be terminated after about 40 sec.

Now let us consider a small time interval Δt. Changes of the volumes of CSF, arterial and venous blood may occur as well as changes of brain volume, for instance produced by a space occupying lesion. Arterial cerebral blood volume is a function of inflow and outflow which can be calculated from pressure differences and the corresponding flow resistances. Assuming the pressure in the venous part of the vascular bed (PV) to be nearly equal the intracranial pressure (ICP) a change of cerebral arterial blood volume during Δt is given by

$$
\Delta VA = \left(\frac{ABP - PA}{C\widehat{V}RI} - \frac{PA - ICP}{CVRA}\right) \cdot \Delta t \ .
$$

Using the compliance (CA) of the arterial part of the vascular bed this volume change can be related to the change of the transmural pressure (PT = PA—ICP):

$$
\Delta VA = CA \cdot (\Delta PA - \Delta ICP) \ ,
$$

where the compliance CA is a function of PT:

$$
CA = \frac{CAN}{1 + (PT/PTN)^4}\left(1 + e^{\frac{PTN}{PT} - 1}\right) \ .
$$

PTN is the transmural pressure under normal conditions (appr. 80 mm Hg), while CAN is the corresponding compliance (appr. 0.06 ml/mm Hg). At normal or moderately elevated ICP the fundamental relation between compliance and transmural pressure

for the arterial part of the cerebral vascular bed is similar to that of the aorta. With increasing ICP, *i.e.* with decreasing transmural pressure the walls of the vessels become flaccid resulting in equal changes of ICP and PA[35]. This requires an unlimited CA when the transmural pressure vanishes, what is ensured by the exponential term in the above equation.

CSF volume is a function of formation and absorption. The absorption rate is defined as the pressure difference between ICP and sinus sagittalis pressure (SSP) divided by the resistance to outflow (RO)[9, 10]. Thus a change of CSF volume can be expressed by

$$
\Delta VCSF = \left(RF - \frac{ICP - SSP}{RO}\right) \cdot \Delta t \ ,
$$

where RF represents the CSF formation rate. This rate is assumed to depend on the regional blood flow through the choroid plexuses:

$$
RF = k \cdot rCBF_{chor.\ plex.} \qquad (k = const.) \ .
$$

Experimental studies of total and regional cerebral blood flow[48] permit to describe the regional flow through the chloroid plexuses as a function of total cerebral blood flow (Fig. 3). Using a polynomial of third degree to fit the experimental data we obtain

$$
RF = 0.36 \cdot CBFR^3 - 0.81 \cdot CBFR^2 + 0.85 \cdot CBFR
$$

with
$$
CBFR = \frac{CBF}{CBFN} \ .
$$

CBFN represents the total cerebral blood flow under normal conditions (750 ml/min). CBF is the actual value given by

$$
CBF = \frac{ABP - ICP}{CVRI + CVRA} \ .
$$

Volume and pressure in the CSF compartment are exponentially related (Fig. 4) with an additive term PO[2, 12, 37, 47]. According to the concept of Sullivan et al.[45] ICP is expressed as

$$
ICP = a \cdot e^{KV \cdot (VCSF - VCSF_{eq})} + PO
$$

where a and KV are constants and $VCSF_{eq}$ represents the equilibrium volume. The additive term PO seems not to be a constant[27, 41]. Assuming $PO = CVP + k$ with a constant k we obtain for the change of ICP

$$
\Delta ICP = KV \cdot (ICP - PO) \cdot \\ \cdot (\Delta VCSF - \Delta VCSF_{eq}) + KP \cdot \Delta CVP \ .
$$

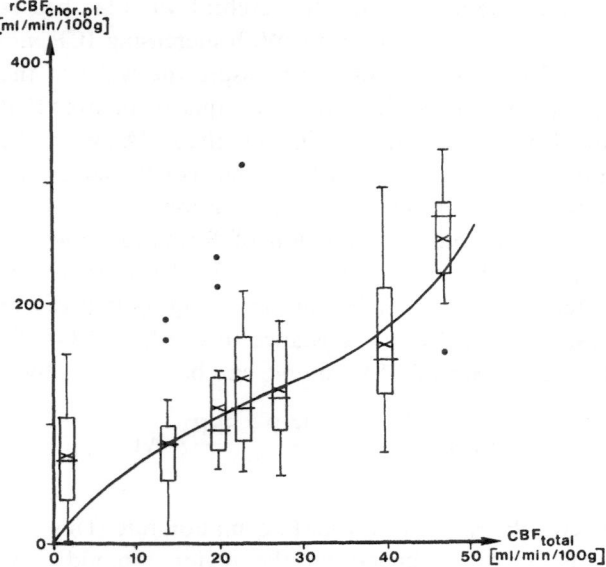

Fig. 3. Regional blood flow through the choroid plexuses in relation to total cerebral blood flow during increasing intracranial pressure. Results from an experimental study in 17 cats[48]

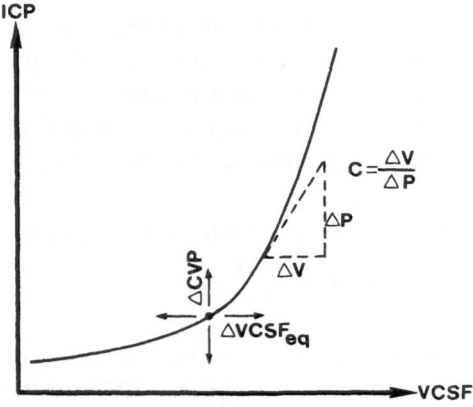

Fig. 4. Exponential volume pressure relationship of CSF space. Changes of equilibrium volume ($VCSF_{eq}$) result in shifting the curve along the volume axis, while changes of central venous pressure cause vertical shifting

An additional space occupying lesion or an increase of cerebral arterial blood volume is compensated by an equivalent decrease of CSF equilibrium volume[45]. Using

$$\Delta VCSF_{eq} = -\Delta VB - \Delta VA$$

the change of ICP can be written as

$$\Delta ICP = KV \cdot (ICP - PO) \cdot \left[\left(RF - \frac{ICP - SSP}{RO} \right) \cdot \Delta t \right.$$
$$+ \Delta VB + \left(\frac{ABP - PA}{CVRI} - \frac{PA - ICP}{CVRA} \right) \cdot \Delta t \right]$$
$$+ KP \cdot \Delta CVP .$$

Defining the compliance of the CSF compartment by

$$CV = \frac{1}{KV \cdot (ICP - PO)} ,$$

dividing the above equations by Δt and replacing the difference quotients by the corresponding differential quotients we obtain the following system of differential equations describing the intracranial CSF and haemodynamics:

$$\frac{dICP}{dt} = KP \cdot \frac{dCVP}{dt} + \frac{1}{CV} \cdot \left(\frac{dVB}{dt} + \frac{ABP - PA}{CVRI} \right.$$
$$\left. - \frac{PA - ICP}{CVRA} + RF - \frac{ICP - SSP}{RO} \right)$$

$$\frac{dPA}{dt} = \frac{dICP}{dt} + \frac{1}{CA} \cdot \left(\frac{ABP - PA}{CVRI} - \frac{PA - ICP}{CVRA} \right)$$

$$\frac{dVA}{dt} = \frac{ABP - PA}{CVRI} - \frac{PA - ICP}{CVRA}$$

$$\frac{dVCSF}{dt} = RF - \frac{ICP - SSP}{RO}$$

$$\frac{dVV}{dt} = -\frac{dVB}{dt} - \frac{dVA}{dt} - \frac{dVCSF}{dt} .$$

The last equation resulting from the Monroe-Kellie-Burrows doctrine can be omitted, because it does not influence the validity of the model. Fig. 5 shows the block diagram of the model of the isolated intracranial system.

2. Cardiovascular Components

The above model permits the study of biomathematical aspects of the ICP dynamics under various conditions. The influence of haemodynamics upon ICP and derived functions can be investigated in detail as demonstrated in the following sections. Effects of increased ICP upon circulation could not be simulated or analysed with this model. For this purpose our model has been extended by inserting it into a simplified model of the cardiovascular system (Fig. 6). The left ventricle of the heart, the aorta and an extracerebral peripheral resistance have been included, while the pulmonary circulation has been omitted.

The left ventricle of the heart is regarded as a rhythmically acting pump moving the stroke volume SV into the aorta during each cardiac cycle. Heart rate (HR) and stroke volume (SV) are not independent. The

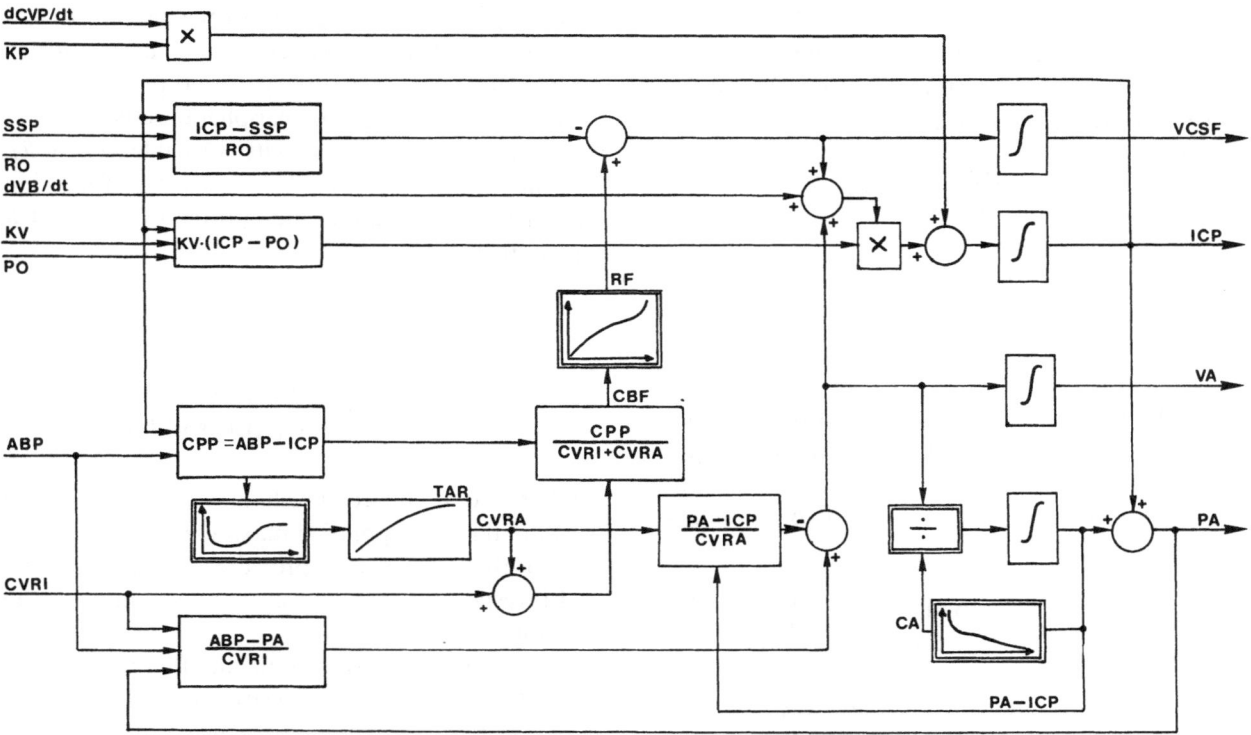

Fig. 5. Block diagram of the mathematical model of the isolated intracranial system

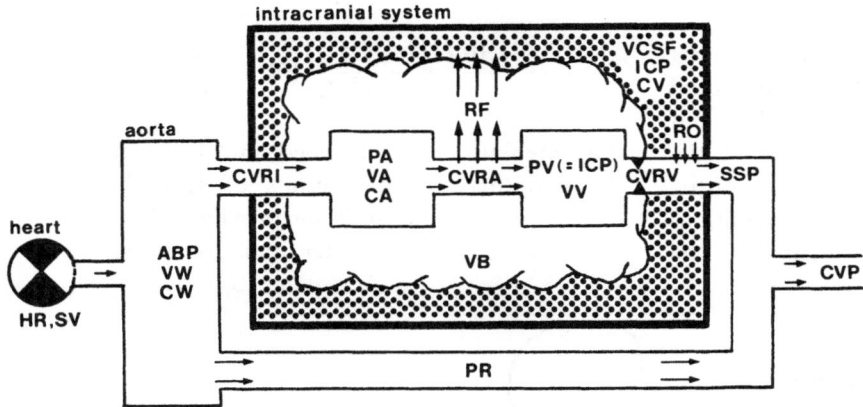

Fig. 6. Extended model of CSF and haemodynamics including cardiovascular
components

stroke volume decreases with increasing heart rate. According to Kirchheim[24] the cardiac output per minute (HMV) is a parabolic function of the heart rate

$$HMV = a \cdot HR - b \cdot HR^2 \ .$$

The coefficients a and b have been calculated assuming the HMV maximum at HR = 180/min and HMV = 5,000 ml/min under normal conditions, *i.e.* HR = 75/min.

The aorta and major arteries are modeled as an elastic reservoir (well known by the German expression "Windkessel") converting the intermittent inflow into a continuous outflow. Via the arterial compliance CW changes of pressure and volume are related by

$$\Delta VW = CW \cdot \Delta ABP \ .$$

The arterial compliance (CW) is a non-linear function of the arterial blood pressure (ABP). Möller et al.[33] used the following equation

$$CW = \frac{2 \cdot CWN}{1 + (ABP/ABPN)^4} \ ,$$

where ABPN is the normal value of ABP (appr. 100 mm Hg) and CWN is the aortic compliance at ABP = ABPN (appr. 0.75 ml/mm Hg). Outflow from the aortic reservoir into the intracranial system can be written as

$$V_{\text{out. i}} = \frac{ABP - PA}{CVRI} \cdot \Delta t \ .$$

For the extacranial part we have

$$V_{\text{out. e}} = \frac{ABP - CVP}{PR} \cdot \Delta t \ ,$$

where CVP represents the central venous pressure and PR denotes the extracerebral peripheral resistance. Introducing differential quotients similar to the procedure described above we obtain the two additional differential equations

$$\frac{dABP}{dt} = \frac{1}{CW} \cdot \left(SV \cdot HR - \frac{ABP - PA}{CVRI} - \frac{ABP - CVP}{PR} \right)$$

$$\frac{dVW}{dt} = SV \cdot HR - \frac{ABP - PA}{CVRI} - \frac{ABP - CVP}{PR} \ .$$

3. Baroreceptor Feedback Control

It has to be noted that heart rate and peripheral resistance are inputs to the extended model and arterial blood pressure is not yet a regulated quantity. To avoid this disadvantage the short-time behaviour of the nervous blood pressure regulation with the aid of the baroreceptor reflex is included into the model additionally. An increase or decrease of ABP sensed by the

baroreceptors causes alterations of the discharge rates in the carotid sinus nerves leading to the medullary centres. Via efferent pathways appropriate inverse changes of heart rate and peripheral resistance are initiated to bring ABP back to its normal level (Fig. 7). According to Möller et al.[33] we use the following equations to approximate the baroreceptor feedback loop characteristics:

$$\widehat{HR} = HRB + \frac{HRM - HRB}{1 + (ABP/ABPN)^8}$$

$$\widehat{PR} = PRB + \frac{PRM - PRB}{1 + (ABP/ABPN)^6} \ .$$

HRB and PRB are threshold values, HRM and PRM are the corresponding maximum values (Fig. 8). ABPN is the arterial blood pressure value at which $\widehat{HR}$ and $\widehat{PR}$ respectively are halfway between minimum and maximum.

The signal transmission along the efferent pathways causes delayed reactions of heart rate and peripheral resistance. By the differential equations

$$\frac{dHR}{dt} = -\frac{1}{TKR} \cdot HR + \frac{1}{TKR} \cdot \widehat{HR}$$

$$\frac{dPR}{dt} = -\frac{1}{TKR} \cdot PR + \frac{1}{TKR} \cdot \widehat{PR}$$

this efferent delay is included into the model. On the basis of experimental results of Koepchen et al.[25] TKR = 5 sec has been chosen providing that a new HR or PR level is reached after appr. 20–30 sec.

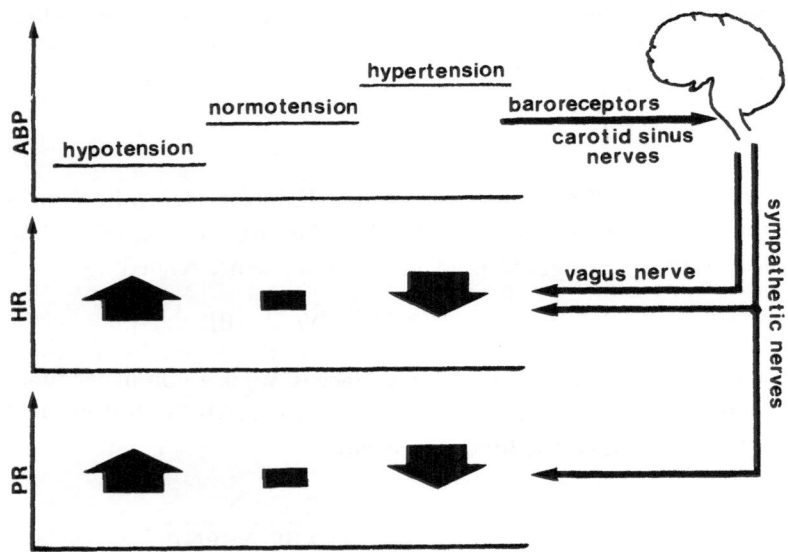

Fig. 7. Sketch of the shorttime regulation of arterial blood pressure with the aid of the baroreceptor reflex. (Modified scheme, according to Rushmer[40])

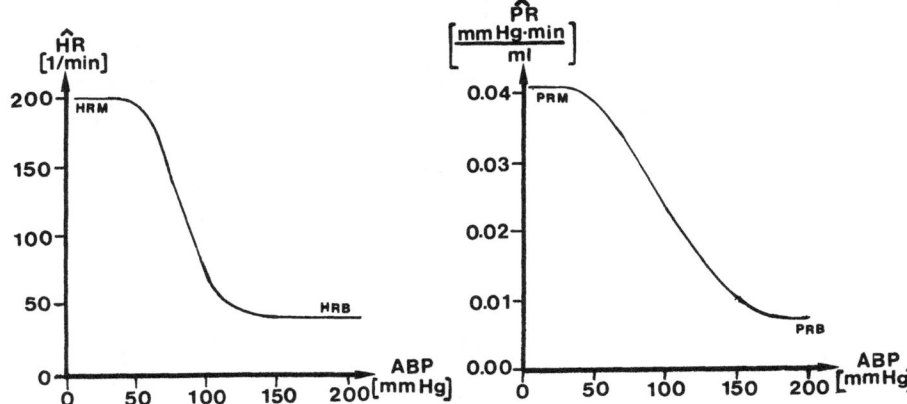

Fig. 8. Baroreceptor feedback loop characteristics. HR and PR as functions of ABP

4. Disturbance of Central Regulation

Acute brain lesions, increased ICP and reduction of cerebral blood flow lead by involvement of the brain stem to disturbances of the central regulation[38, 39]. Since quantitative measurements of central dysfunctions directly transferable into a mathematical model are not known our approach has been chosen as simple as possible.

The actual value of the regional blood flow through the brain stem in relation to the normal flow value has been regarded as a measure of the disturbance of the ABP control circuit. Similar to the regional blood flow through the choroid plexuses we can express the brain stem flow as a function of the total cerebral blood flow (Fig. 9):

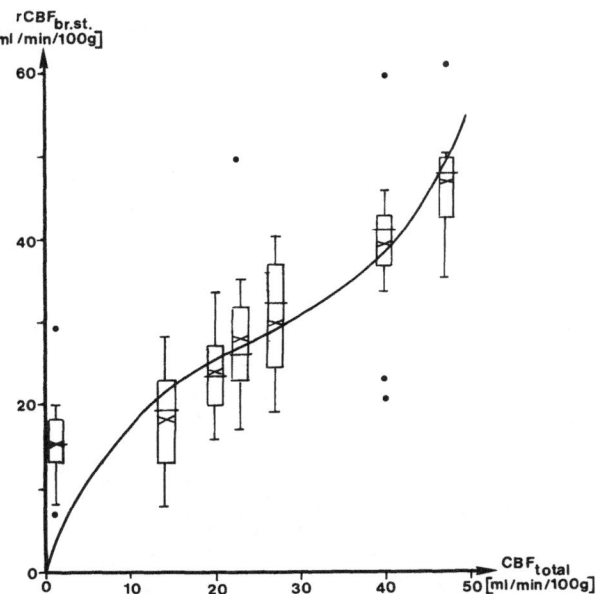

Fig. 9. Regional blood flow through the brain stem in relation to total cerebral blood flow during increasing intracranial pressure. Results from an experimental study in 17 cats[48]

$$\frac{rCBF_{br.\,st.}}{CBFN} = 1.27 \cdot CBFR^3 - 2.82 \cdot CBFR^2 + 2.55 \cdot CBFR \; .$$

where

$$CBFR = \frac{CBF}{CBFN} \; .$$

Now we define a quantity $\hat{DRF}$ expressing the degree of brain stem dysfunction by

$$\hat{DRF} = \begin{cases} 1 & \frac{rCBF_{br.\,st.}}{CBFN} \geqslant LD \\[2ex] \dfrac{rCBF_{br.\,st.}}{LD \cdot CBFN} & \frac{rCBF_{br.\,st.}}{CBFN} < LD \; . \end{cases}$$

The delimiter LD, its value has to be near below one, prevents dysregulation in the case of only small reduction of brain stem flow. Furthermore the differential equation

$$\frac{dDRF}{dt} = -\frac{1}{TDR} \cdot DRF + \frac{1}{TDR} \cdot \hat{DRF}$$

ensures that shortlasting flow reductions result in moderate disturbances of regulation. The effect of brain stem dysfunction upon the ABP control circuit is then expressed by application of a modified argument to the transition functions:

$$\hat{HR} = \hat{HR} \, (DRF^{NHR} \cdot ABP)$$

$$\hat{PR} = \hat{PR} \, (DRF^{NPR} \cdot ABP) \; .$$

Due to the possibility that brain stem dysfunction may influence heart rate reaction and peripheral resistance reaction in different ways the powers NHR and NPR have been introduced.

An additional factor possibly affecting the brain stem function may be the rapidity of an ICP increase.

With decreasing ICP this factor shall be inoperative. Taking in account a delayed disturbance of restorage of brain stem function we introduce the variable STG by the differential equation

$$\frac{dSTG}{dt} = -\frac{1}{TST} \cdot STG + \frac{1}{TST} \cdot \widehat{STG}$$

where $\widehat{STG}$ is given by

$$STG = \begin{cases} \dfrac{dICP}{dt} & \dfrac{dICP}{dt} \geqslant 0 \\[2ex] 0 & \dfrac{dICP}{dt} < 0 \end{cases}$$

Furthermore we define a factor STF by

$$STF = 1 - \frac{1 - STFB}{1 + \dfrac{STG}{2}} \qquad 0 < STFB < 1$$

varying between the basic value STFB and one. Finally we use this factor to modify the above derived factor of dysregulation via

$$DRF = 1 - STF \cdot (1 - DRF) \ .$$

5. Space Occupying Lesions

It is assumed that an expanding space occupying lesion will affect the resistance to outflow (RO) via cisternal obstruction and blockade of CSF pathways. With RON as the resistance to outflow under normal conditions and with VBN as the normal volume of brain tissue we define the actual resistance to outflow by

$$\frac{RON}{RO} = \begin{cases} 1 & VB \leqslant VBL \\[1ex] 1 - \dfrac{VB - VBL}{VBM - VBL} & VBL < VB < VBM \\[1ex] 0 & VB \geqslant VBM \ . \end{cases}$$

Below a limit VBL the CSF absorption is not affected. A further increase of VB will cause an elevation of RO leading to total cessation of CSF absorption when VB exceeds VBM.

A block diagram of the extended model including vascular components, short-time regulation of ABP, space occupying lesions and brain stem dysfunction is shown in Fig. 10.

Stability, Model Validation and Simulation Technique

The local stability of this non-linear model has been tested by linearization about the equilibrium points[18].

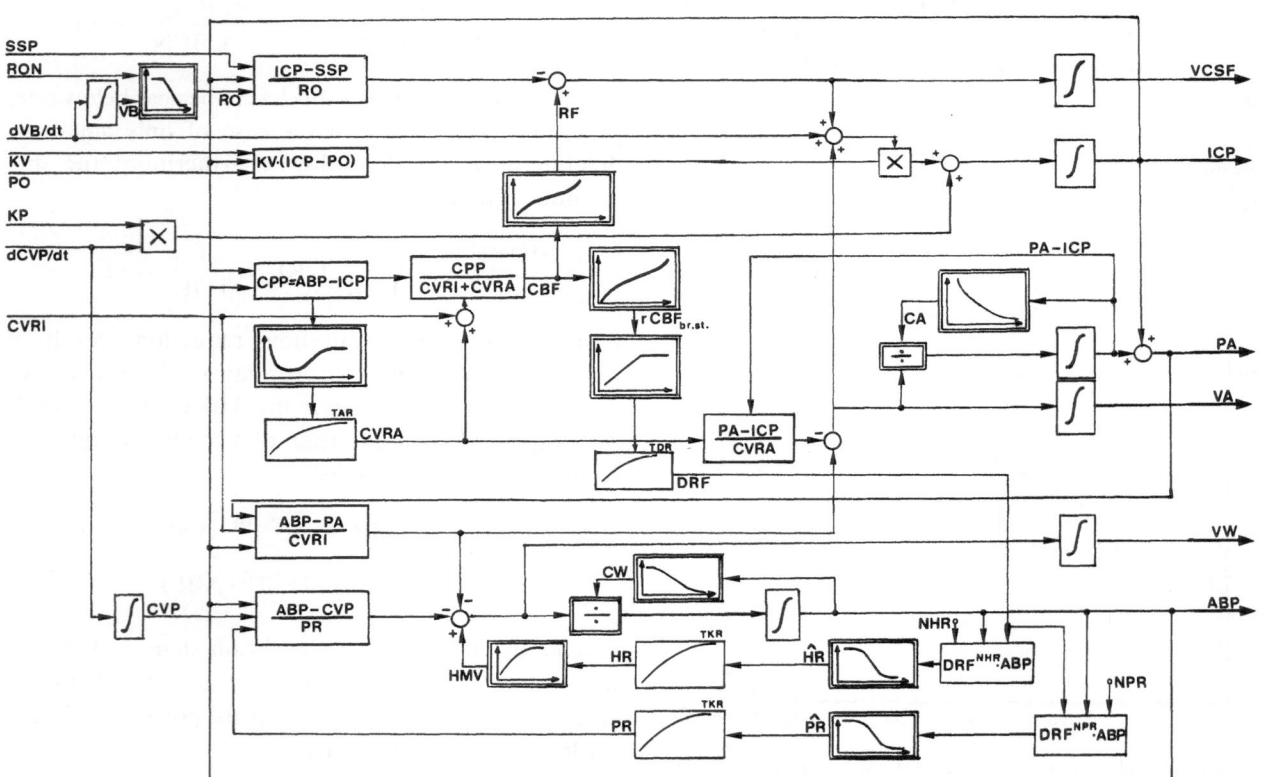

Fig. 10. Block diagram of the extended model

This method is applicable only when the system has been slightly displaced away from an equilibrium point. The behaviour of the model in the case of greater disturbances has been studied by simulation.

Model validation has been performed comparing the dynamic behaviour of the model and that of the real system under various conditions, *i.e.* volume pressure test, expanding space occupying lesions with different growth rates and intracranial pulse pressure relationship during increasing ICP. Since the model is represented by a set of nonlinear differential equations the existence of a closed solution in terms of elementary functions seems to be unlikely. Therefore the Runge Kutta method of 4th order[8] has been applied for simulation. For the time step usually $\Delta t = 1/960$ min has been used, while for simulation of pulsatile behaviour $\Delta t = 1/3,840$ min has been needed.

Model Applications

In general a mathematical model offers three possibilities of application:

— Theoretical derivations from the basic formulas describing the model.

— Parameter estimation by adjusting model parameters to fit model output and measured data optimally.

— Simulation studies to investigate the system behaviour under different conditions.

The applicability of our model with respect to these topics is demonstrated in the following sections.

1. Intracranial Pulse Pressure Relationship and Haemodynamics

The relation of intracranial pressure amplitude (AICP) to the actual mean intracranial pressure (MICP)—called further pulse pressure relationship (PPR)—is influenced by many factors. Because of the haemodynamic origin of the ICP amplitude it can be expected that arterial blood pressure (ABP) and its amplitude (AABP) affect the intracranial pulse pressure relationship.

To study these influences by simulation the model of the isolated intracranial space has been used. The arterial blood pressure as the input to the system has been approximated by an appropriate Fourier sum with ten terms[18]. Isolated variations of MABP and AABP were used to examine the effect of these parameters upon the PPR (Fig. 11). PPR has been calculated by producing different MICP levels via alteration of the resistance to outflow (RO) and by computing the time courses over several seconds.

This analysis revealed an opposite effect of MABP and AABP upon the PPR. At a constant MABP the slope of PPR becomes steeper with increasing AABP. When AABP is kept constant the intracranial pressure amplitude decreases with increasing MABP. Additionally certain deviations from a basically linear PPR has been revealed by these simulations. For a detailed assessment of these phenomena we apply a pure sine wave as ABP input. Furthermore we assume the

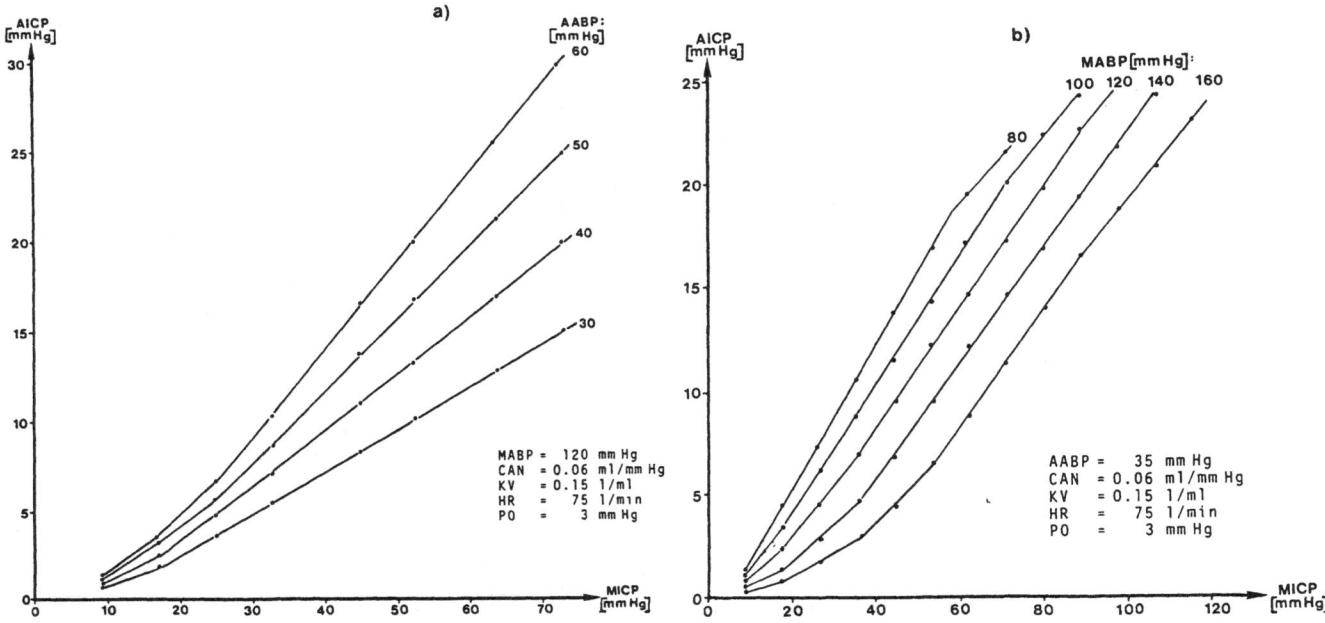

Fig. 11. Influence of the arterial blood pressure amplitude (a) and of the mean arterial blood pressure (b) upon the intracranial pulse pressure relationship

variable coefficients CV, CA, CVRA and RF to be constant over one cardiac cycle. Since in this case the differential equations possess closed solutions in terms of trigonometric functions we are able to calculate the transfer function

$$U = \frac{AICP}{AABP}$$

After elimination of negligible terms[18] we obtain

$$\frac{AICP}{AABP} \simeq$$

$$\frac{1}{\sqrt{\left(1 + \frac{CV}{CA}\left(1 + \frac{CVRI}{CVRA}\right)\right)^2 + (2\pi HR \cdot CV \cdot CVRI)^2}}.$$

Using this equation the influence of KV (determining the intracranial compliance CV) and of CAN (determining the compliance CA of the arterial part of the cerebral vascular bed) upon the transfer function U can be analysed (Fig. 12). Marked deviations from a linear relation between U and MICP are observed. When CAN or KV becomes smaller one or two breakpoints occur.

The above transfer function U contains two competitive terms under the square root sign. At high MICP levels, i.e. at low transmural pressures the compliance of the arterial vascular bed will increase rapidly, so that the transfer function is essentially determined by the

second term. Thus the lack or occurrence of a breakpoint depends on the relation between the compliance of the CSF space and that of the arterial cerebral vascular bed.

2. Volume Pressure Test and Haemodynamics

The influence of the arterial blood pressure upon the intracranial pressure increase resulting from a volume pressure test has been analysed by simulation with the model of the isolated intracranial system. For different levels of ABP and ICP rapid injection of 1 ml of artificial CSF during 1 sec has been simulated. ICP response (ΔICP) as a function of the initial MICP for different ABP levels is shown in Fig. 13.

According to the exponential volume pressure relationship of the CSF space the pressure response depends linearly on the initial level when MICP does not exceed 30 mm Hg. Below this limit the pressure response is more pronounced with an elevated ABP causing an underestimation of intracranial compliance.

This error remains below 7%. Since this accuracy usually is not given in real measurements an insignificant effect of ABP upon the volume pressure test can be postulated.

When MICP exceeds the limit of about 30 mm Hg the volume pressure response flattens or decreases. The occurrence of this effect is positively correlated with ABP.

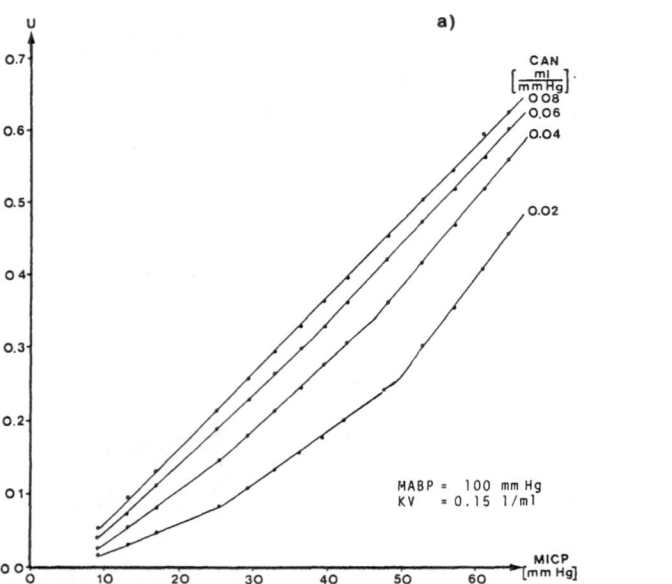
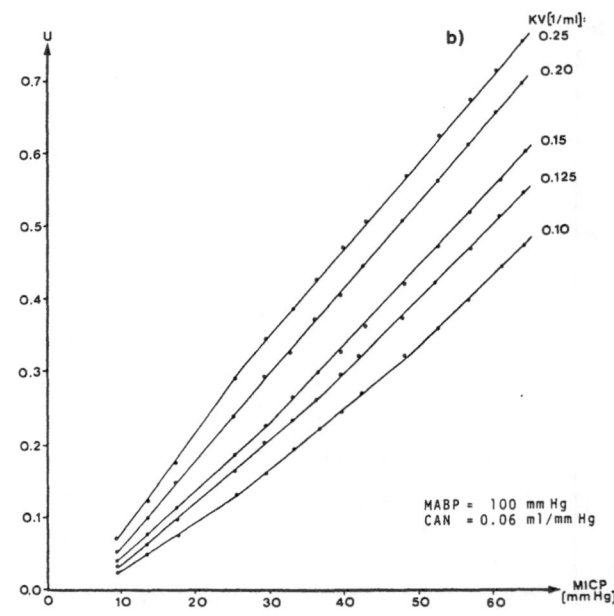

Fig. 12. Influence of the parameters CAN (determining compliance of the arterial part of the cerebral vascular bed) and of KV (determining compliance of the CSF space) upon the transfer function U (= ICP amplitude/ABP amplitude)

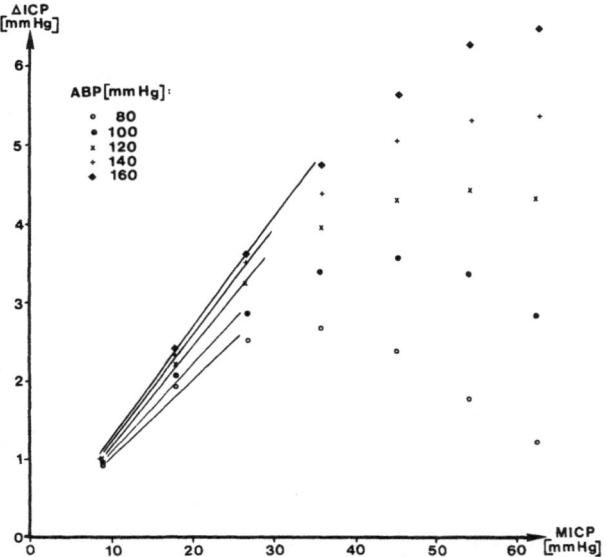

Fig. 13. Influence of arterial blood pressure upon the pressure response (ΔICP) to a volume pressure test

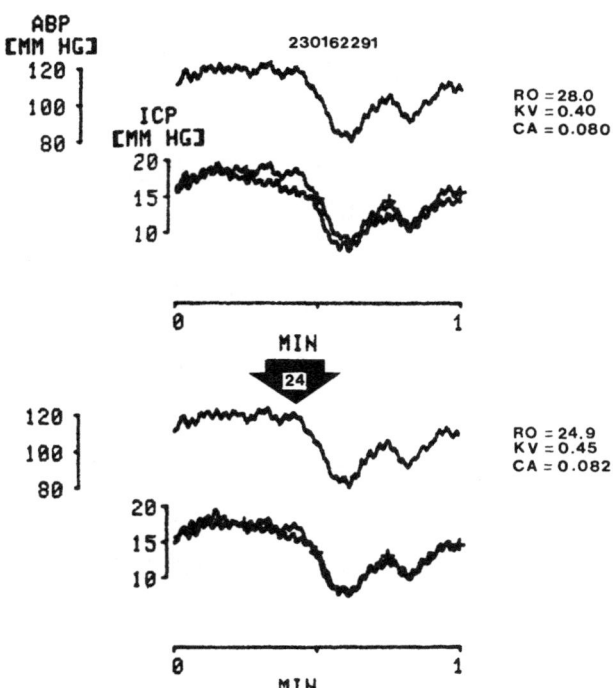

Fig. 14. Parameter estimation applied to a spontaneous decerebration episode over one minute. Optimal coincidence of measured and simulated (+) data after 24 iterations

3. Parameter Estimation

Parameter estimation means the evaluation of unknown and directly not measurable parameters of the system. This task can be solved by adjusting the model parameters in such a way that the difference between the outputs of the model and of the real system becomes minimal. A summary of different identification methods and demonstrations of practical applications has been given by Isermann[21] and by Möller et al.[33].

In our study we have applied parameter estimations to the model of the isolated intracranial system using the measured ABP as input. As the error criterion which has to be minimized we have chosen

$$MQA = \frac{1}{N} \cdot \sum_1^N (ICP_m - ICP_s)^2 \to Min \ ,$$

where N is the number of measurements within the analysed time interval. ICP_m denotes the measured values, while ICP_s are the data obtained by simulation.

Fig. 14 demonstrates simultaneous identification of the parameters RO, KV and CAN in the case of a spontaneous decerebration episode over one minute. The initial parameter estimates already result in an acceptable coincidence between measured and simulated ICP providing MQA = 1.1 mm Hg. By systematic search in the three-dimensional space spanned by the above parameters the deviation has been reduced to MQA = 0.04 mm Hg after 24 iterations.

4. Rhythmic Phenomena

Central dysregulation due to primary or secondary brain stem lesions is often accompanied by synchronous reactions and rhythms of vegetative variables[28, 38, 39]. To simulate these phenomena the extended model has been used. For this simulation study the initial levels of ICP have been elevated so that cerebral blood flow has been already reduced. In all examples with different parameter sets the same pattern of rhythmic reactions has been observed: Synchronous rapid increase of ABP, HR and ICP similar to those occuring in patients with decerebration (Fig. 15).

Variation of the parameters NHR and NPR influences the power of the reaction or the amplitude of the rhythm but not the rhythm itself. Alteration of the time constant TST related to the restoration time after a disturbance of the brain stem function directly affects the periodicity of this reaction.

Discussion

The application of mathematical methods to biomedical problems has been proven to be advantageous in various situations. Corresponding to advances in med-

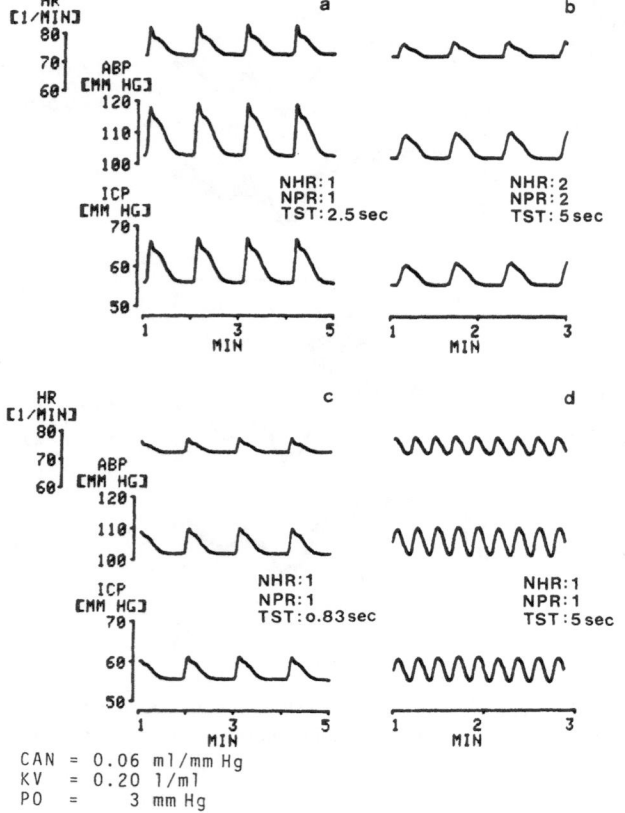

CAN = 0.06 ml/mm Hg
KV = 0.20 l/ml
PO =   3 mm Hg

Fig. 15. Simulations of disturbances of central regulation of ABP with different parameter sets. Rhythmic occurrence of simultaneous increase of ICP, ABP and HR

ical sciences more complex statistical procedures and simulation models have been required. This has been supported by the dynamic progress in electronic data processing[42, 43].

Because of their complex structure the model of the isolated intracranial system as well as the extended model can not be compared with already established ones. None of the known models includes simultaneously a flow dependent CSF formation, autoregulation of cerebral blood flow, cerebral haemodynamics and cardiovascular elements. The cardiovascular system has been modeled with great simplifications though more detailed models exist[33]. Our approach seems to be reasonable taking into account that a model will not be improved automatically by simple accumulation of subsystems. Furthermore the main task was to study the dynamics of the intracranial system and its interrelations with the cardiovascular system, but not the cardiovascular system itself.

It is obvious that our model maps only a limited segment of the real system into appropriate mathematical equations. Metabolic and respiratory factors, for instance, have been omitted. This has to be taken into account when interpreting simulation results. Furthermore this model does not permit studies concerning the morphology of the CSF pulse wave. Because of the "Windkessel" like structure of the cerebral vascular bed only informations refering to mean pressures and amplitudes can be deduced from this model. When pulsatile effects and the morphology of the pulse wave are of interest one has to apply the mathematics of wave theory[7, 23].

For discussion of the analysis of haemodynamic effects upon the intracranial pulse pressure relationship only few literature sources are available. Szewczykowski et al.[46] reported a breakpoint in PPR at small MICP levels defining a low elastance zone with a pressure independent ICP amplitude. The influence of arterial blood pressure and its amplitude upon PPR has already been predicted with the aid of a simpler provisional model and has been confirmed in a set of clinical data by partial correlation analysis including MICP, AICP, MABP and AABP[20]. These results are supported by experimental findings of Avezaat and van Eijndhoven[2]. Additionally they have observed changes of PPR in terms of division into two linear functions similar to those obtained by our simulations under certain conditions. These authors concluded that a breakpoint indicates a loss of CBF autoregulation. In contrast to their conclusion the theoretical derivation of the transfer function from our model equations suggests the breakpoint as an effect of the relation between the compliance of the CSF space and that of the arterial part of the cerebral vascular bed.

The decreased or flattened ICP response to a volume pressure test is at first surprising, but the same effect has been observed by Avezaat and van Eijndhoven in an experimental series of six dogs[2]. From the breakpoints in the PPR and in the ICP response to a volume pressure test these authors deduced a linear pressure volume curve of the CSF space at ICP levels above the breakpoint. On the other hand our model acting with an exponential pressure volume curve over the whole ICP range has shown the same behaviour as observed in animal experiments. This demonstrates at least that the observation of a small number of variables does not permit generally valid conclusions concerning the structure of a dynamic system such as the intracranial space.

Many attempts have been made to evaluate parameters of the intracranial system, in particular intracranial elastance or compliance, from data obtained during clinical monitoring of ICP only. These procedures deriving the intracranial compliance from the pulse pressure relationship need additional volume pressure tests for calibration[3, 11]. Other methods have been described to calculate intracranial elastance directly from PPR without any further impairment of the patient[14, 46]. However, the latter approaches seem to be insufficient because they are based exclusively upon PPR disregarding the haemodynamic influences.

The application of our model to parameter identification of the intracranial system is based on modern concepts of systems analysis. These procedures offer the advantage to estimate not only a single value but a set of different system parameters. However, some limitations concerning the method as well as our actual state of its implementation on a digital computer are to be noticed. Only those segments of a recording are suitable for application of parameter estimation where variables not included into the model or considered as constant remain stable or do not change significantly. Furthermore detailed studies about the identifiability of the parameters and about the consistency of the estimates are lacking.

The simulation program has been written in a higher language (Fortran). Because of the small time step needed for sufficient accuracy a simulation run consumes as much time as the real recording. This means, for example, that an iterative parameter estimation with twenty-four iterations needs twenty-four times the real recording time of the analysed segment. In future this may be improved by using assembler language for programming and by application of identification procedures with a faster convergence.

The approach to simulation of rhythmic phenomena in central dysregulation generates patterns of reaction as recorded during intensive care monitoring described in detail by Lorenz[28]. The simulation demonstrated in the present paper shall give an outlook to future models including cerebrospinal fluid system, cardiovascular system and neurodynamical system[19]. The interpretation and classification of rhythmicity of vegetative variables in brain stem lesions may possibly be supported by the extended model. Application of parameter identification techniques to the extended model may enable quantitative description of those phenomena by a small set of data facilitating inter- and intraindividual comparative research.

References

1. Agarwal GC, Berman BM, Stark L (1969) A lumped parameter model of the cerebrospinal fluid system. IEEE Trans Biomed Eng 16: 45–53

2. Avezaat CJJ, Eijndhoven JHM van (1984) Cerebrospinal fluid pulse pressure and craniospinal dynamics. A theoretical, clinical and experimental study. Thesis, Erasmus University, Rotterdam

3. Avezaat CJJ, Eijndhoven JHM van, Jong DA de, Moolenaar WCJ (1976) A new method of monitoring intracranial volume-pressure relationship. In: Beks JWF, Bosch DA, Brock M (eds) Intracranial pressure III. Springer, Berlin Heidelberg New York, pp 308–313

4. Benabid AL, Rougemont J de, Barge M (1985) CSF dynamics: A mathematical approach. In: Lundberg N, Pontén U, Brock M (eds) Intracranial pressure II. Springer, Berlin Heidelberg New York, pp 54–60

5. Bloch R, Talalla A (1976) A mathematical model of cerebrospinal fluid dynamics. J Neurol Sci 27: 485–498

6. Buslenko NP (1972) Modellierung komplizierter Systeme. Verlag Die Wirtschaft, Berlin

7. Busse R, Bauer RD (1982) Arteriensystem. In: Busse R (ed) Kreislaufphysiologie. Thieme, Stuttgart New York, pp 41–69

8. Collatz L (1966) The numerical treatment of differential equations. Springer, Berlin Heidelberg New York

9. Cutler RWP, Page L, Galicich J, Watters GV (1968) Formation and absorption of cerebrospinal fluid in man. Brain 91: 707–720

10. Davson H, Hollingsworth G, Segal MB (1970) The mechanism of drainage of the cerebrospinal fluid. Brain 93: 665–678

11. Eijndhoven JHM van, Avezaat CJJ, Wyper DJ (1980) The CSF pulse pressure in relation to intracranial elastance and failure of autoregulation. In: Shulman K, Marmarou A, Miller JD, Becker DP, Hochwald GM, Brock M (eds) Intracranial pressure IV. Springer, Berlin Heidelberg New York, pp 153–158

12. Eijndhoven JHM van, Sliwka S, Avezaat CJJ (1986) The constant pressure term (Po) of the volume-pressure relationship. Comparison between results of infusion test and pulse pressure analysis. In: Miller JD, Teasdale GM, Rowan JO, Galbraith SL, Mendelow AD (eds) Intracranial pressure VI. Springer, Berlin Heidelberg New York Tokyo, pp 48–53

13. Emshoff JR, Sisson RL (1972) Simulation mit dem Computer. Verlag Moderne Industrie, München

14. Godin D, Stevenaert A, Lhommel R (1980) Study of the CSF pulsation transfer: Application to the frequency analysis. In: Shulman K, Marmarou A, Miller JD, Becker DP, Hochwald GM, Brock M (eds) Intracranial pressure IV. Springer, Berlin Heidelberg New York, pp 191–194

15. Guinane JE (1972) An equivalent circuit analysis of cerebrospinal fluid hydrodynamics. Am J Physiol 223: 425–430

16. Hakim S, Venegas JG, Burton JD (1976) The physics of the cranial cavity, hydrocephalus and normal pressure hydrocephalus. Surg Neurol 5: 187–210

17. Hofferberth B, Matakas F, Fritschka E (1975) A computer model of CSF dynamics. In: Lundberg N, Pontén U, Brock M (eds) Intracranial pressure II. Springer, Berlin Heidelberg New York, pp 61–66

18. Hoffmann O (1985) Ein mathematisches Modell zur Simulation und Analyse der intrakraniellen Liquor- und Hämodynamik. Eine medizinisch-theoretische Studie. Habilitationsschrift, Gießen

19. Hoffmann O (1987) Some aspects of the application of neurodynamical models for the simulation of central regulation and dysregulation. In: Möller DPF (ed) System analysis of biological processes. Vieweg, Braunschweig Wiesbaden, pp 188–193

20. Hoffmann O, Zierski JT (1982) Analysis of the ICP pulse-pressure relationship as a function of arterial blood pressure.

Clinical validation of a mathematical model. Acta Neurochir (Wien) 66: 1–21

21. Isermann R (1972) Methoden zur Identifikation von mathematischen Modellen für das dynamische Verhalten biologischer Systeme. Biomed Techn 17: 218–227

22. Johnston IH, Rowan JO (1974) Raised intracranial pressure and cerebral blood flow. 3. Venous outflow tract pressures and vascular resistances in experimental intracranial hypertension. J Neurol Neurosurg Psychiat 37: 392–402

23. Kenner T, Ronninger R (1960) Untersuchung über die Entstehung der normalen Pulsform. Arch Kreislaufforsch 32: 141–173

24. Kirchheim H (1982) Kreislaufregulation. In: Busse R (ed) Kreislaufphysiologie. Thieme, Stuttgart New York, pp 167–210

25. Koepchen HP, Lux HD, Wagner PH (1961) Untersuchungen über Zeitbedarf und zentrale Verarbeitung des pressorezeptorischen Herzreflexes. Pflügers Archiv 273: 413–430

26. Löfgren J (1973) Effects of variations in arterial pressure and arterial carbon dioxide tension on the cerebrospinal fluid pressure-volume relationships. Acta Neurol Scand 49: 586–598

27. Löfgren J, Essen C von, Zwetnow NN (1973) The pressure-volume curve of the cerebrospinal fluid space in dogs. Acta Neurol Scand 49: 557–574

28. Lorenz R (1973) Wirkungen intrakranieller raumfordernder Prozesse auf den Verlauf von Blutdruck und Pulsfrequenz. Acta Neurochir (Wien) [Suppl 20], Springer, Wien New York

29. Lowell HM, Bloor BM (1971) The effect of increased intracranial pressure on cerebrovascular haemodynamics. J Neurosurg 34: 760–768

30. Marmarou A, Shulman K, Rosende RM (1978) A nonlinear analysis of the cerebrospinal fluid system and intracranial pressure dynamics. J Neurosurg 48: 332–344

31. May RM (1976) Simple mathematical models with very complicated dynamics. Nature 261: 459–467

32. McIntosh JEA, McIntosh RP (1980) Mathematical modelling and computers in endocrinology. Springer, Berlin Heidelberg New York

33. Möller D, Popović D, Thiele G (1983) Modeling, simulation and parameter-estimation of the human cardiovascular system. In: Hartman I (ed) Advances in control systems and signal processing, vol 4. Vieweg & Sohn, Braunschweig Wiesbaden

34. Niemeyer G (1977) Kybernetische System- und Modelltheorie. System dynamics. Franz Vahlen, München

35. Nornes H, Aaslid R, Lindegaard KF (1977) Intracranial pulse

pressure dynamics in patients with intracranial hypertension. Acta Neurochir (Wien) 38: 177–186

36. Ommaya AK, Metz H, Post KD (1972) Observations on the mechanics of hydrocephalus. In: Harbert JC (ed) Cisternography and hydrocephalus. Ch C Thomas, Springfield, pp 57–74

37. Paltsev EIP, Sirovsky EB (1982) Intracranial physiology and biomechanics. Clinical data on pressure-volume relationships and their interpretation. J Neurosurg 57: 500–510

38. Pia HW (1973) Central dysregulation. Z Neurol 204: 1–21

39. Pia HW (1985) Primary and secondary lesions of hypothalamus and brain stem. Adv Neurosurg 13: 217–253

40. Rushmer RF (1976) Cardiovascular dynamics. WB Saunders, Philadelphia London Toronto

41. Ryder HW, Espey FF, Kimbell FD, Penka EJ, Rosenauer A, Podolsky B, Evans JP (1953) The mechanism of the change in cerebrospinal fluid pressure following an induced change in the volume of the fluid space. J Lab Clin Med 41: 428–435

42. Schneider B (1967) Über Simulation biologischer Modelle. Elektromedizin 12: 3–9

43. Seelos HJ (1979) Biomedizinische Simulationsmodelle. EDV Med Biol 10: 97–101

44. Spain JD (1982) Basic microcomputer models in biology. Addison-Wesley, London Amsterdam Don Mills Ontario Syndney Tokyo

45. Sullivan HG, Miller JD, Searle JR (1980) An interpretation of pressure/volume interactions in the craniospinal axis. Neurosurgery 6: 453–462

46. Szewczykowski J, Sliwka S, Kunicki A, Dytko P, Korsak-Sliwka J (1977) A fast method of estimating the elastance of the intracranial system. A practical application in neurosurgery. J Neurosurg 47: 19–26

47. Tans JTJ, Poortvliet DCJ (1986) Comparison of pressure volume indices obtained with constant rate and bolus infusions. In: Miller JD, Teasdale GM, Rowan JO, Galbraith SL, Mendelow AD (eds) Intracranial pressure VI. Springer, Berlin Heidelberg New York Tokyo, pp 79–83

48. Zierski J, Kurzaj E, Hoffmann O, Winkler B (1983) Cerebral blood flow in the brain stem during increased ICP. In: Ishii S, Nagai H, Brock M (eds) Intracranial pressure V. Springer, Berlin Heidelberg New York Tokyo, pp 452–457

Author's address: PD Dr. Oskar Hoffmann, Department of Neurosurgery, University of Giessen, Klinikstrasse 29, D-6300 Giessen, Federal Republic of Germany.

Satz und Druck: Adolf Holzhausens Nfg., Universitätsbuchdrucker

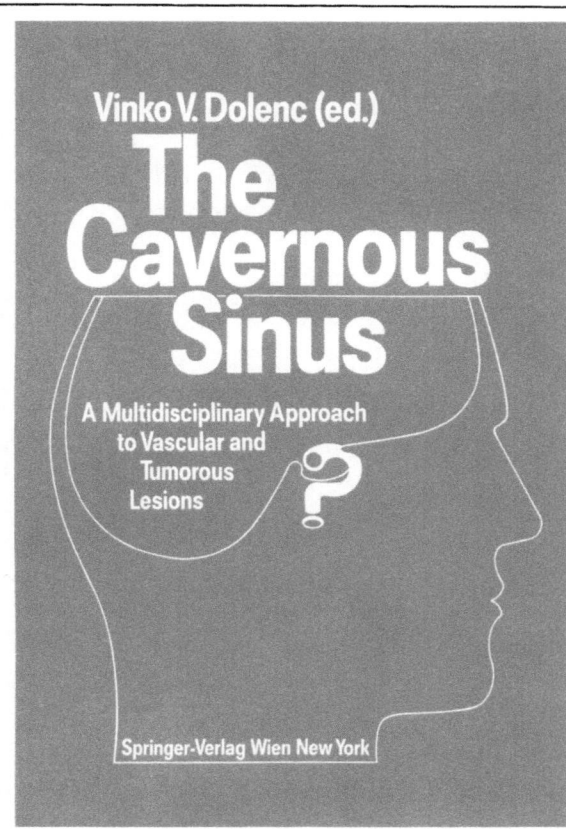

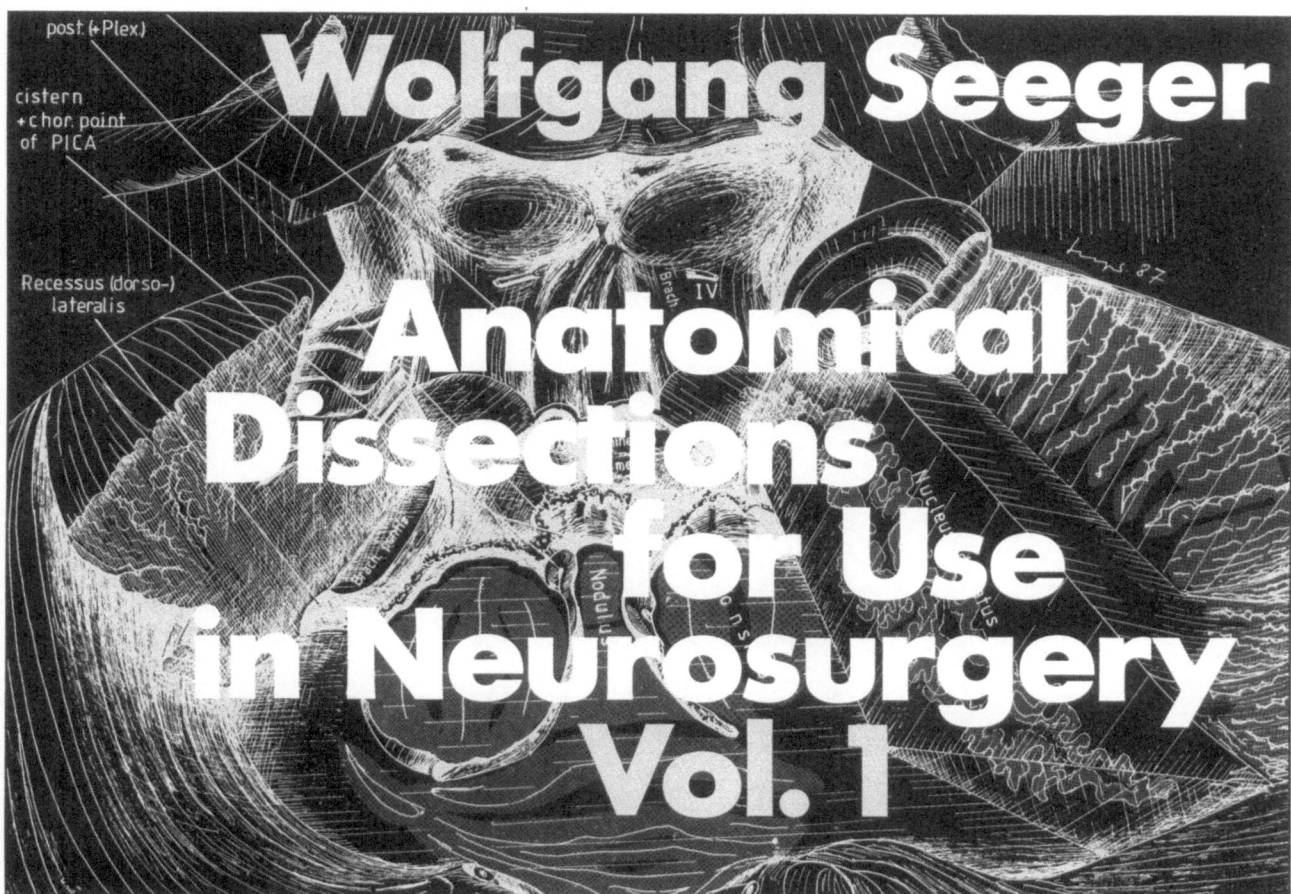

post. (+Plex.)

cistern
+chor. point
of PICA

Recessus (dorso-)
lateralis

IV

Wolfgang Seeger
Anatomical Dissections for Use in Neurosurgery Vol. 1

Advances in imaging techniques and microsurgery have made it possible to operate on any intracerebral structure. However, precise knowledge of the anatomy of the brain is an essential prerequisite.

Conventional textbooks of the anatomy of the brain present the individual systems according to their developmental and functional aspects, e. g cortex, limbic system, basal ganglia, brainstem and cerebellum. As during the operation the neurosurgeon may be confronted with all the various systems, it is necessary to perform brain dissections under the aspect of operative approaches.

The first volume gives young neurosurgeons methodical instructions in the dissection of the formaldehyde-preserved brain. Certain individual surface structures are kept intact until the end of the dissection so that the relative topographical positions of surface and deep structures can be made discernable. Two specimens are needed: In the first dissection the cerebrum and cerebellum are studied to make the young neurosurgeon familiar with these structures during his initial training. The blood vessels and leptomeninges are removed to make the structures more easily recognizable and less confusing. As experience has shown that by the end of the first dissection the specimen is rather damaged by mistakes, the relationship between deep seated parts, the ventricular system and surface structures are then shown in a second dissection, where the vessels and leptomeninges are preserved.

Volume 2
(in preparation): The second volume, scheduled
for publication in 1988, is intended for experienced neurosurgeons
who perform complex operations deep inside the brain.
Before such operations, which have not become standardized,
it is advisable (the author still does
so occasionally) to orientate oneself
on a formaldehyde-preserved brain.

Volume 1
1987. 150 figures.
IX, 313 pages.
Cloth DM 228,–, öS 1600,–
ISBN 3-211-81998-3

SPRINGER-VERLAG WIEN NEW YORK

Moelkerbastei 5, A-1010 Wien ● Heidelberger Platz 3, D-1000 Berlin 33 ● 175 Fifth Avenue, New York, NY 10010, USA ● 37-3, Hongo 3-chome, Bunkyo-ku, Tokyo 113, Japan